Advanced Maths Essentials Mechanics 1 for OCR

T0187325

Welcome to Advanced Maths Essentials: Mechanics 1 for OCR. This book will help you to improve your examination performance by focusing on all the essential maths skills you will need in your OCR Mechanics 1 exam. It has been divided by chapter into the main topics that need to be studied. Each chapter has then been divided by sub-headings, and the description below each sub-heading gives the OCR specification for that aspect of the topic.

The book contains scores of worked examples, each with clearly set-out steps to help solve the problem. You can then apply the steps to solve the Skills Check questions in the book and past exam questions at the end of each chapter. If you feel you need extra practice on any topic, you can try the Skills Check Extra exercises on the accompanying CD-ROM. At the back of this book there is a sample exam-style paper to help you test yourself before the big day.

Some of the questions in the book have a ⊚ symbol next to them. These questions have a PowerPoint® solution (on the CD-ROM) that guides you through suggested steps in solving the problem and setting out your answer clearly.

Using the CD-ROM

To use the accompanying CD-ROM simply put the disc in your CD-ROM drive, and the menu should appear automatically. If it doesn't automatically run on your PC:
1. Select the My Computer icon on your desktop.
2. Select the CD-ROM drive icon.
3. Select Open.
4. Select mechanics1_for _ocr.exe.

If you don't have PowerPoint® on your computer you can download PowerPoint 2003 Viewer®. This will allow you to view and print the presentations. Download the viewer from http://www.microsoft.com

Pearson Education Limited
80 Strand
London
WC2R 0RL
England
www.longman.co.uk

First published 2005
11
ISBN 978-0-582-83661-7

Design by Ken Vail Graphic Design

Cover design by Raven Design

Typeset by Tech-Set, Gateshead

Printed in China (CTPS/11)

The publisher wishes to draw attention to the Single-User Licence Agreement at the back of the book. Please read this agreement carefully before installing and using the CD-ROM.

We are grateful for permission from OCR to reproduce past exam questions. All such questions have a reference in the margin. OCR can accept no responsibility whatsoever for accuracy of any solutions or answers to these questions.

Every effort has been made to ensure that the structure and level of sample question papers matches the current specification requirements and that solutions are accurate. However, the publisher can accept no responsibility whatsoever for accuracy of any solutions or answers to these questions. Any such solutions or answers may not necessarily constitute all possible solutions.

1 Mathematical models in mechanics

Mathematical models can be used in mechanics to make predictions about real-life situations. Models must be tested and improved by comparing their predictions against experimental data, but once they successfully describe the situation they can then be used to describe many similar situations.

In order to develop a model you must simplify the real-life situation by making **modelling assumptions.** For example, if you are trying to model a car that travels 5 km, you may ask where you treat as the starting point of measurement. Do you measure from the front end of the car or the back end? Because the car is small relative to the distance travelled, you may assume that the car can be treated as a particle. This then becomes a modelling assumption. Similarly, the time taken for the car to travel may be influenced by wind; it will reach a given destination faster if there is a strong wind travelling in its direction of travel. In M1 you assume that any effects due to wind can be neglected. There are several terms and definitions used in M1 that you should know (you may be examined on these types of modelling assumptions):

Objects treated as particles: in general, it is assumed that objects that are small relative to the other sizes involved will be treated as particles; this means that the mass of the object can be considered to act at the single point where the particle is placed, e.g. an aeroplane can be modelled as a point particle, as it travels long distances compared to its length.

Lamina: this is a flat object whose thickness can be ignored, as the thickness is small compared to the other dimensions (its width and its length). For example, a thin sheet of card may be represented as a lamina.

Rigid body: this is an object that is made up of different parts or masses which do not move relative to each other. For example, a car can be considered to be a rigid body, i.e. when the engine produces a forward force, the whole car moves.

Rod: this is an object in which the mass is considered to concentrate along a line; it is assumed to have only length, and not breadth nor width.

Uniform object (lamina or rod): if an object is uniform then the mass is evenly distributed across the object and can be considered to act at its centre. For example, you could make the assumption that a tree trunk is a uniform rod and that its mass can be considered to act from its centre. However, this may not be a reasonable modelling assumption if the base of the trunk is thicker than the top of the trunk. In this case, the trunk is non-uniform and the centre of mass may have to be calculated.

Light object: when the mass of an object is small compared to the rest of the masses involved, it is said to be a light object. For example, when modelling the mass of the Earth, the mass of a car is relatively light, so the mass of the car may be ignored in the model.

Light and inextensible strings: in M1, strings are assumed to be:

1 light, i.e. the mass of the string is small relative to the other masses involved. This is a fair assumption, unless you are using a chain instead of a string, in which case the mass may have to be included.

2 inextensible, i.e. you don't have to account for any force to extend the string. Again, this is a fair assumption, as the extension in a string is relatively small.

More complicated problems, without these assumptions, are introduced in M3.

Smooth surface: when a surface is assumed to be smooth then you assume that there is no resistive force opposing the movement due to the contact of the surface with the object, i.e. you assume that there is no **friction**. For example, when modelling an ice rink, you may assume that the surface is smooth.

Rough surface: when the surface is not smooth then it is said to be rough. In this case the friction must be included when modelling the situation. For example, sand paper being rubbed across a brick will have a comparatively large force of friction that cannot be ignored because the contact between the surfaces is rough.

Smooth and light pulleys: in M1, pulleys are assumed to be:

1 smooth, i.e. there is no friction on the surface or the bearings of the pulley.

2 light, i.e. the mass of the pulley is considered to be small in comparison to the other masses involved.

Bead: this is a particle that can be threaded onto a string or wire.

Wire: this is a rigid body that is a thin thread of metal.

Peg: this is a support from which an object can be hung or on which an object may rest. It acts as a point but may be either rough or smooth.

Gravity: this is the force that attracts objects towards each other. Because of the relatively massive size of the Earth, it can be assumed that objects only experience attraction towards the Earth and not towards each other. The force of gravity reduces in size as you move further away from the surface of the Earth. However, in all the models in M1, gravity is assumed to be constant.

Air resistance: this is a force that is experienced because of the resistance of the air. For example, when you drop a sheet of paper, the paper will not fall as fast as a brick. However, in M1 it is generally assumed that there is no air resistance.

Wind: this is a force that can be felt because of the action of the wind. In M1 it is assumed that there is no wind.

2 Force as a vector

2.1 Vectors and forces

Understand the vector nature of a force, and use directed line segments to represent force (acting in at most two dimensions).

Vectors

A **vector quantity** is one that has both size and direction, as opposed to a **scalar quantity** that has size only. When you add two scalar quantities, you can add the numbers directly. When you add two vector quantities, you must take their directions into account first.

Example 2.1

Step 1: Draw a clear diagram, marking all known angles and distances.

A man walks 2 km due north from point O to point A and then 3 km due east to point B. Find the distance OB and the bearing of the point B from O.

 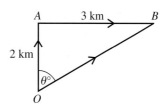

Step 2: Use trigonometry to calculate unknown angles and distances.

$$OB^2 = OA^2 + AB^2$$
$$= 2^2 + 3^2 = 13$$
$$OB = \sqrt{13} \text{ km}$$
$$\tan \theta° = \tfrac{3}{2}$$
$$\theta = 56.3\ldots$$

The bearing is 056° (nearest degree).

Example 2.2

The point A is on a bearing of 045° and at a distance of 25 m from the point O. The point B is on a bearing of 160° and at a distance of 40 m from the point A. Find the distance OB and the bearing of B from O.

Step 1: Draw a clear diagram, marking all known angles and distances.

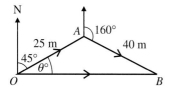

Angle $OAB = 360° - 135° - 160°$
$$= 65°$$
Let angle AOB be θ.

Step 2: Use trigonometry to calculate unknown angles and distances.

$$OB^2 = OA^2 + AB^2 - 2(OA)(AB) \cos 65°$$
$$= 25^2 + 40^2 - 2(25)(40) \cos 65° = 1379.76\ldots$$
$$OB = 37.1 \text{ m (3 s.f.)}$$

$$\frac{\sin \theta°}{AB} = \frac{\sin 65°}{OB}$$

$$\sin \theta° = \frac{40 \sin 65°}{37.1\ldots} = 0.9759\ldots$$

$$\theta = 77.4\ldots$$

The bearing of B from O is $(45° + 77.4\ldots°) = 122°$ (nearest degree).

Notation: In the above example the **distance** OA is 25 m, while the **vector** OA is 25 m on a bearing of 045°. To distinguish between the distance OA and the vector OA an arrow is placed over the letters to represent the vector thus: $\overrightarrow{OA}$.

Example 2.2 can be described in this notation as:

$$\overrightarrow{OB} = \overrightarrow{OA} + \overrightarrow{AB}$$

The vector $\overrightarrow{OB}$ is called the **resultant** of the vectors $\overrightarrow{OA}$ and $\overrightarrow{AB}$.

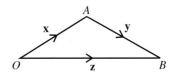

Notation: You can also represent vectors using a single letter in bold type. If the vector $\overrightarrow{OA} = \mathbf{x}$, the vector $\overrightarrow{AB} = \mathbf{y}$, and the vector $\overrightarrow{OB} = \mathbf{z}$, then:

$$\mathbf{z} = \mathbf{x} + \mathbf{y}.$$

Note:
If the vector from O to $A = \mathbf{x}$, then the vector from A to $O = -\mathbf{x}$
i.e. $\overrightarrow{OA} = \mathbf{x}$
$\overrightarrow{AO} = -\mathbf{x}$

The size of the vector $\mathbf{x}$ can be written as $|\mathbf{x}|$ or simply x without the bold type. This refers to the **scalar magnitude**: the distance OA.

If two vectors, $\mathbf{x}$ and $\mathbf{y}$, are equal then you can write:

$$\mathbf{x} = \mathbf{y}$$

Note:
This means that they both have the same size and direction.

If two vectors, $\mathbf{x}$ and $\mathbf{y}$, are parallel but have different magnitudes, then you can write:

$$\mathbf{x} = k\mathbf{y},$$

where k is a scalar.

Note:
This means that they both have the same direction, but not necessarily the same size: the vector $\mathbf{x}$ is k times bigger than the vector $\mathbf{y}$.

When you handwrite vectors, since you cannot show bold type, you underline the letter, for example $\underline{a}$ or $\underset{\sim}{a}$.

Example 2.3 $OABC$ is a parallelogram with $\overrightarrow{OA} = \mathbf{a}$ and $\overrightarrow{OC} = \mathbf{c}$. The point M is the midpoint of BC and the point N is on AB such that $AN = \frac{2}{3}AB$.

Express, in terms of $\mathbf{a}$ and $\mathbf{c}$, the following vectors:

   a $\overrightarrow{AB}$ **b** $\overrightarrow{OB}$ **c** $\overrightarrow{AO}$ **d** $\overrightarrow{BO}$

   e $\overrightarrow{OM}$ **f** $\overrightarrow{CN}$ **g** $\overrightarrow{MN}$

Step 1: Draw a clear diagram, marking all known vectors.

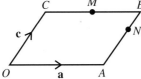

Note:
Because $OABC$ is a parallelogram, BC is parallel to OA and so $\overrightarrow{CB} = \mathbf{a}$ and similarly $\overrightarrow{AB} = \mathbf{c}$.

a $\overrightarrow{AB} = \mathbf{c}$

b $\overrightarrow{OB} = \overrightarrow{OA} + \overrightarrow{AB} = \mathbf{a} + \mathbf{c}$

c $\overrightarrow{AO} = -\mathbf{a}$

d $\overrightarrow{BO} = -\overrightarrow{OB} = -\mathbf{a} - \mathbf{c}$

e $\overrightarrow{OM} = \overrightarrow{OC} + \overrightarrow{CM}$

      $= \overrightarrow{OC} + \frac{1}{2}\overrightarrow{CB}$

      $= \mathbf{c} + \frac{1}{2}\mathbf{a}$

f $\overrightarrow{CN} = \overrightarrow{CB} + \overrightarrow{BN}$

      $= \overrightarrow{CB} + \frac{1}{3}\overrightarrow{BA}$

      $= \mathbf{a} - \frac{1}{3}\mathbf{c}$

g $\overrightarrow{MN} = \overrightarrow{MB} + \overrightarrow{BN}$

      $= \frac{1}{2}\mathbf{a} - \frac{1}{3}\mathbf{c}$

Note:
c $\overrightarrow{AO}$ is in the opposite direction to $\overrightarrow{OA}$.

Note:
f You can also calculate
$\overrightarrow{CN} = \overrightarrow{CO} + \overrightarrow{OA} + \overrightarrow{AN}$
to give the same result.

2.2 Resultant of forces

Understand the term 'resultant' as applied to two or more forces acting at a point, and use vector addition in solving problems involving resultants and components of forces.

Forces

A force is a vector quantity (it has both magnitude and direction). So you can apply the previous techniques to forces. The exact definition of a force, measured in newtons (N), is given in Chapter 5. Forces can also be represented as directed line segments.

If more than one force acts at a point then you can find the overall effect of the forces by finding the **resultant force**. This can be done by drawing a **parallelogram of forces**.

Suppose two forces P N and Q N act at a point O as shown:

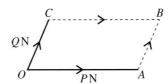

These forces are represented as directed line segments $\overrightarrow{OA}$ and $\overrightarrow{OC}$. A parallelogram is completed by drawing the lines AB and BC. To find the resultant, you have to consider:

$$\overrightarrow{OA} + \overrightarrow{OC}$$

But $\overrightarrow{OC} = \overrightarrow{AB}$ because these vectors are equivalent (they both have the same size and direction). So,

$$\overrightarrow{OA} + \overrightarrow{OC} = \overrightarrow{OA} + \overrightarrow{AB}$$

And so, the resultant of the two forces that are represented by the line segment $\overrightarrow{OA}$ and $\overrightarrow{OC}$ is also fully represented by the line segment $\overrightarrow{OB}$. This is the diagonal of the parallelogram.

One way of finding the resultant forces is by drawing a parallelogram of forces to scale. However, it is more accurate to use trigonometry to find the resultant force.

Resultant of a horizontal and vertical force

A horizontal force F_h N and a vertical force F_v N act at a point as shown. The magnitude of the resultant can be calculated using

$$|\mathbf{F}| = \sqrt{(F_h^2 + F_v^2)}$$

and the angle made with the horizontal component can be calculated using

$$\tan \theta^\circ = \frac{F_v}{F_h}.$$

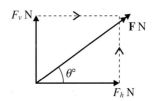

Example 2.4   A force of 8 N acts horizontally to the right and a force of 3 N acts vertically upwards. Find the magnitude of the resultant of these forces and the angle that the resultant makes with the horizontal force.

Step 1: Draw a parallelogram of forces.

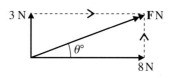

Step 2: Find the resultant force.

$$|F| = \sqrt{(F_h{}^2 + F_v{}^2)}$$
$$= \sqrt{8^2 + 3^2}$$

$$\tan \theta° = \frac{F_v}{F_h} = \frac{3}{8}$$

Step 3: Solve for unknowns.

$$= 8.54 \ldots$$

$$\theta = \tan^{-1}\left(\frac{3}{8}\right) = 20.6\ldots$$

The magnitude of the resultant force is 8.54 N (3 s.f.) and the angle is 20.6° (3 s.f.).

Note:
You may be asked to find the angle that the resultant makes with the upward vertical. This is $90° - 20.6° = 69.4°$.

Resultant of forces that are not perpendicular

If the forces are not perpendicular to each other then the resultant force can be found by making use of the sine and cosine rules.

Example 2.5   Two forces of 3 N and 5 N act at a point such that the angle between them is 55°. Find the resultant force and the angle that the resultant force makes with the 5 N force.

Note:
Interior angles add up to 180°.

Let $\theta°$ be the angle that the resultant makes with the 5 N force.

Step 1: Draw a parallelogram of forces.

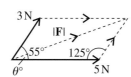

Step 2: Find the resultant force.

Using the cosine rule:

$$|\mathbf{F}|^2 = 3^2 + 5^2 - 2(3)(5) \cos 125°$$

Using the sine rule:

$$\frac{\sin 125°}{|\mathbf{F}|} = \frac{\sin \theta°}{3}$$

Step 3: Solve for unknowns.

$$|\mathbf{F}| = 7.1559\ldots \text{ N}$$

$$\sin \theta° = 0.343\ldots$$
$$\theta = 20.08\ldots$$

The resultant force has magnitude 7.2 N (2 s.f.) and the angle that it makes with the 5 N force is 20° (2 s.f.).

2.3 Resolving a force into components

Find and use perpendicular components of a force, e.g., in finding a resultant of a system of forces, or to calculate the magnitude and direction of a force.

If a force **F** N has magnitude F N and it acts at an angle $\theta°$ to the positive x-axis you can work out its horizontal and vertical components using simple trigonometry:

$$F_h = F \cos \theta°$$

$$F_v = F \sin \theta°$$

Note:
Define $\theta°$ as the angle between the force and the positive x-axis.

Example 2.6 Find the horizontal and vertical components for a force of 10 N, which acts at an angle of $140°$ to the positive x-axis.

Step 1: Draw a force diagram resolving each force into its components.

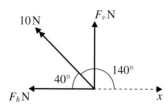

Tip:
Use the acute angle to work out the components, then incorporate a sign to give the direction.

Step 2: Find the components in each direction.

Horizontal component
$= 10 \cos 40°$ to the left
so $F_h = -7.66$ (3 s.f.)

Vertical component
$= 10 \sin 40°$ upwards
so $F_v = 6.43$ (3 s.f.)

Note:
The resultant
$\sqrt{((-7.66...)^2 + 6.43...^2)} = 10$

Resultant of two or more forces

The resultant of two or more forces can be calculated by summing the vertical and horizontal components, separately.

Example 2.7 Find the resultant and the angle that the resultant makes with the 5 N force for the following system of forces.

Step 1: Draw the force diagram, and, where appropriate, redraw each force resolved into any two perpendicular directions.

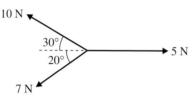

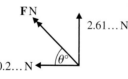

Recall:
Horizontal and vertical components are perpendicular to each other. A horizontal force has 0 component vertically.

Note:
Vectors can be resolved into any two perpendicular directions.

Step 2: Find the resultant in each direction.

Horizontal component:
$F_h = 5 - 10 \cos 30° - 7 \cos 20°$
$\quad = -10.23...$

Vertical component:
$F_v = 10 \sin 30° - 7 \sin 20°$
$\quad = 2.60...$

Note:
Remember to use opposite signs for forces in opposite directions.

$F = \sqrt{(F_h^2 + F_v^2)} = 10.5$ (3 s.f.)

Step 3: Find the overall resultant and the angle.

$\theta = \tan^{-1}\left(\dfrac{F_v}{|F_h|}\right) = 14.3$ (3 s.f.)

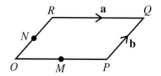

The resultant is a force of 10.5 N at an angle of $(180° - 14.3...°) = 166°$ (3 s.f.) with the 5 N force.

SKILLS CHECK **2A: Force as a vector**

1 A man walks x km from O on a bearing of $\theta°$ to a point A. He then walks a further distance of y km on a bearing of $\alpha°$ to a point B. Find the distance OB and the bearing of O from B when:

a $\theta = 090$, $\alpha = 180$, $x = 4$ and $y = 3$ **b** $\theta = 024$, $\alpha = 160$, $x = 7$ and $y = 2$

c $\theta = 290$, $\alpha = 045$, $x = 5$ and $y = 2$

2 The diagram shows a parallelogram $OPQR$. The vector $\mathbf{a}$ represents $\overrightarrow{RQ}$ and the vector $\mathbf{b}$ represents $\overrightarrow{PQ}$. M is the midpoint of OP and N is the midpoint of OR.

Find, in terms of $\mathbf{a}$ and $\mathbf{b}$, the following vectors:

a $\overrightarrow{OQ}$ **b** $\overrightarrow{OP}$ **c** $\overrightarrow{OM}$

d $\overrightarrow{MQ}$ **e** $\overrightarrow{QN}$ **f** $\overrightarrow{MN}$

3 The diagram shows a triangle OAB with $\overrightarrow{OA}$ = **a** and $\overrightarrow{OB}$ = **b**.

The point C is on AB such that $\overrightarrow{AC}$ = $3\overrightarrow{CB}$.

Express the following vectors in terms of **a** and **b**:

a $\overrightarrow{AB}$ **b** $\overrightarrow{BA}$ **c** $\overrightarrow{AC}$ **d** $\overrightarrow{OC}$

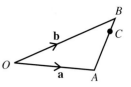

4 Find the magnitude of the resultant and the angle that the resultant makes with the positive x-axis for the following forces, giving your answers to one decimal place:

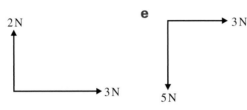

a 6 N, 3 N

b 5 N, 5 N

c 10 N, 18 N

d 2 N, 3 N

e 3 N, 5 N

5 Find the horizontal and vertical components of the following forces:

a a force of 10 N acting at 10° to the horizontal

b a force of 25 N acting at 15° to the horizontal

c a force of 35 N acting at 85° to the horizontal.

6 Find the magnitude of the horizontal and vertical components of the following forces that act on a particle, evaluated to one decimal place:

a a force of 5 N acting at 120° to the horizontal

b a force of 20 N acting at 30° below the horizontal

c a force of 145 N acting at 220° to the horizontal.

7 Find the resultant of the following forces:

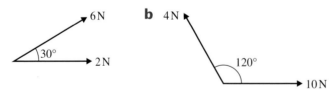

a 6 N, 30°, 2 N

b 4 N, 120°, 10 N

8 The diagram shows forces **P** N, **Q** N and **R** N acting on a particle.

The line of action of the force **Q** is in the horizontal direction.

Find the magnitude of the resultant of these forces and the angle that the resultant makes with the direction of force **Q** if:

a **P** = 10 acting at 30° above the horizontal, **Q** = 7 and **R** = 15 acting at 15° below the horizontal

b **P** = 50 acting at 19° above the horizontal, **Q** = 45 and **R** = 100 acting at 18° below the horizontal.

SKILLS CHECK **2A EXTRA** is on the CD

1

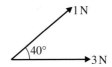

Two forces, of magnitude 1 N and 3 N, act on a particle in the directions shown in the diagram. Calculate the magnitude of the resultant force on the particle and the angle between this resultant force and the force of magnitude 3 N.   [OCR Specimen]

2

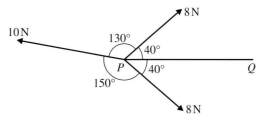

Fig. 1   **Fig. 2**

Forces of magnitudes 8 N and 5 N act on a particle. The angle between the directions of the two forces is 30°, as shown in Fig. 1. The resultant of the two forces has magnitude R N and acts at an angle θ° to the force of magnitude 8 N, as shown in Fig. 2. Find R and θ.   [OCR June 2002]

 3

Three forces, of magnitudes 8 N, 10 N and 8 N, act at a point P in the direction shown in the diagram. PQ is the bisector of the acute angle between the two forces of magnitude 8 N. Find

i   the components of the resultant of the three forces

    a   parallel to PQ,

    b   perpendicular to PQ,

ii   the magnitude of the resultant of the three forces,

iii the angle that the resultant of the three forces makes with PQ.   [OCR June 2003]

4 Two forces act at a point along a smooth horizontal plane. The forces have magnitudes 12 N and 15 N and the angle between the forces is 120°. Find

    a   the resultant of the two forces

    b   the angle that the resultant makes with the 15 N force.   [OCR Nov 2003]

 5 Two forces act in a vertical plane. The forces have magnitudes 20 N and 7 N and make angles α and β respectively with the upward vertical, as shown in the diagram. The angles α and β are such that $\cos \alpha \approx 0.96$ and $\cos \beta \approx 0.6$. [You are given that $\sin \alpha \approx 0.28$ and $\sin \beta \approx 0.8$.]

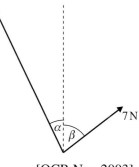

    i   Show that the resultant of the two forces acts vertically and find its magnitude.

    ii   The two forces act on a particle of mass 2.5 kg. State, giving a reason, whether the direction of the acceleration of the particle is vertically upwards or vertically downwards.

[OCR Nov 2003]

3 Equilibrium of a particle

3.1 Equilibrium

Understand and use the principle that a particle is in equilibrium if and only if the vector sum of the forces acting upon it is zero, or equivalently if and only if the sum of the resolved parts in any direction is zero.

A system of forces acting on a particle is in equilibrium when the resultant force is zero (there is no net force).

- **The algebraic sum of the horizontal components is 0 N.**

- **The algebraic sum of the vertical components is 0 N.**

In fact, a system of forces acting on a particle is in equilibrium if the sum of the resolved parts of the force in any direction is 0 N.

Example 3.1   Find P and Q if the following system of forces is in equilibrium:

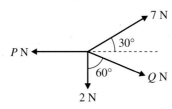

Recall:
$F_h = F\cos\theta°$ and $F_v = F\sin\theta°$ where $\theta°$ is defined as the angle between the vector and the positive x-axis. The angle that Q makes with the vertical is 60° so the angle Q makes with the horizontal is 30°.

Step 1: Draw the force diagram, and, where appropriate, redraw each of the forces resolved into any two perpendicular directions.

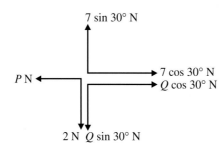

Note:
Define up and right as positive, down and left as negative.

Step 2: Find the resultant force in each direction (and equate each to zero when in equilibrium).

Step 3: Solve for unknowns.

Vertical components:

$7\sin 30° - 2 - Q\sin 30° = 0$

$Q = \dfrac{7\sin 30° - 2}{\sin 30°} = 3$

Force Q is 3 N.

Horizontal components:

$7\cos 30° + Q\cos 30° - P = 0$

$P = 7\cos 30° + 3\cos 30°$

$= 8.660\ldots$

Force P is 8.66 N (3 s.f.).

Example 3.2   The following diagram shows a system of forces acting on a particle in a plane. A third force is added so that the particle rests in equilibrium. Find the magnitude of this force and the angle that it makes with the horizontal.

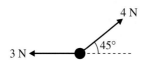

Step 1: Draw the force diagram, and, where appropriate, redraw each of the forces resolved into any two perpendicular directions.

Let the force which is added to keep the system in equilibrium be F N and let the angle it makes below the horizontal be $\theta°$.

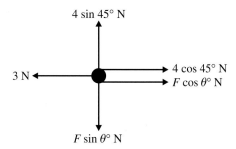

Step 2: Find the resultant force in each direction (and equate each to zero when in equilibrium).

Step 3: Solve for unknowns.

Horizontal components:

$4 \cos 45° + F \cos \theta° - 3 = 0$

$F \cos \theta° = 3 - 4 \cos 45°$ (1)

Vertical components:

$4 \sin 45° - F \sin \theta° = 0$

$F \sin \theta° = 4 \sin 45°$ (2)

Divide (2) by (1)

$$\tan \theta° = \frac{\sin \theta°}{\cos \theta°} = \frac{4 \sin 45°}{3 - 4 \cos 45°} = 16.48\ldots$$

$$\theta = \tan^{-1} 16.48\ldots$$

$$= 86.5 \ (3 \text{ s.f.})$$

$$F = \frac{4 \sin 45°}{\sin 86.5°}$$

$$= 2.83 \ (3 \text{ s.f.})$$

The added force is 2.83 N, acting at an angle of 86.5° below the horizontal.

3.2 Types of force

Identify forces acting in a given situation, and use the relationship between mass and weight; use Newton's third law; use the model of smooth contact.

Weight is the gravitational attraction between a particle of mass m kg and the Earth. It is often written as:

$$\text{weight} = mg$$

where g is the acceleration due to gravity and has approximate value 9.8 m s^{-2}. Weight is a force that always acts vertically downwards, towards the centre of the Earth.

Newton's third law states that every action has an equal and opposite reaction, i.e. if one particle applies a force on another particle, the other particle applies an equal force on the first but in the opposite direction. This is the principle behind the normal reaction force.

Normal reaction force is the force exerted by a surface on a particle (the surface exerts an equal and opposite force on the particle that rests on the surface) which lies in equilibrium. This force always acts in a direction that is perpendicular to the surface, opposing the direction of the weight. This is a consequence of Newton's third law.

Tension is the resisting force provided by a string, when the string holds the particle in equilibrium. It acts away from the particle.

Thrust is the resisting force provided by a spring, when the spring holds the particle in equilibrium. It acts in a direction to oppose the compression or extension.

Friction is the force that opposes motion on a **rough** surface. It is caused by the contact between the particle and the surface. On a **smooth** surface the friction is 0 N.

These forces are used in unit M1. You can apply the techniques learnt above to questions involving these forces acting on a particle.

Example 3.3 A block of mass 3 kg rests on a rough, horizontal surface. The block is pushed with a horizontal force of 3 N. The block is kept in equilibrium by a frictional force F N. Find F and the normal reaction.

Step 1: Draw the force diagram and, where appropriate, redraw each of the forces resolved into any two perpendicular directions.

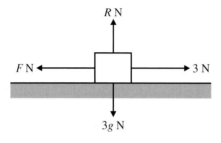

Note:
You can draw a pushing force on one side as a pulling force on the opposite side.

Recall:
The frictional force will act to oppose the pushing force of 3 N.

Step 2: Find the resultant force in each direction (and equate each to 0 when in equilibrium).

Horizontal components:

$F - 3 = 0$

$F = 3$

Vertical components:

$R - 3g = 0$

$R = 3g$

$= 29.4$

Note:
Equate the horizontal and vertical components to zero separately.

Step 3: Solve for unknowns.

The frictional force is 3 N.

The normal reaction is 29.4 N.

Recall:
$g = 9.8$

Sometimes, it may be assumed that the surface is smooth i.e. there is a smooth contact between a particle and the surface that it rests upon. This is a fair modelling assumption in certain cases (for example, if the particle rests on ice). However, this is only a modelling assumption (see Chapter 1) and has limitations because, to some degree, friction is always present. In the case of a smooth contact, it is assumed that friction is 0 N.

Example 3.4 A particle of mass 3 kg rests on a smooth horizontal table. A string with tension T N acts at an angle of 45° above the horizontal and pulls the particle. Another force of 15 N acts at an angle of 30° below the horizontal and opposes the tension in the string to keep the particle in equilibrium. Find the reaction force and the tension in the string.

Step 1: Draw the force diagram and, where appropriate, redraw each of the forces resolved into any two perpendicular directions.

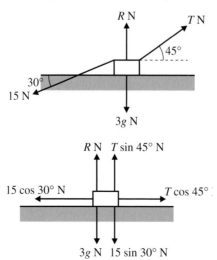

Note:
The method is the same as in the previous section. The only change is in describing the unknowns.

Step 2: Find the resultant force in each direction (and equate each to 0 when in equilibrium).

Step 3: Solve for unknowns.

Horizontal components:

$T \cos 45° - 15 \cos 30° = 0$

$T = \dfrac{15 \cos 30°}{\cos 45°} = 18.37\ldots$

The tension in the string is 18.4 N (3 s.f.).

Vertical components:

$R + T \sin 45° - 15 \sin 30° - 3g = 0$

$R = 15 \sin 30° + 3g - T \sin 45°$
$= 23.90\ldots$

The normal reaction is 23.9 N (3 s.f.).

Sometimes a system of forces may act to suspend a particle freely in the air. In this case there will be no reaction force because the particle is not in contact with any surface.

Example 3.5 A particle of weight 19.6 N is suspended freely by two strings and hangs in equilibrium as shown. Find the tension in the two strings.

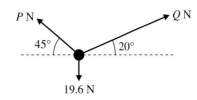

Step 1: Draw the force diagram and, where appropriate, redraw each of the forces resolved into any two perpendicular directions.

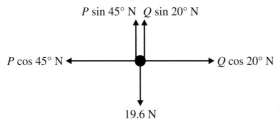

Note: Remember to include the weight.

Step 2: Find the resultant force in each direction (and equate each to 0 when in equilibrium).

Step 3: Solve for unknowns.

Horizontal components:

$Q \cos 20° - P \cos 45° = 0$

$Q = \dfrac{P \cos 45°}{\cos 20°}$

Substitute $P = 20.32\ldots$

$Q = 15.29\ldots$

Vertical components:

$P \sin 45° + Q \sin 20° - 19.6 = 0$

$19.6 = P \sin 45° + \dfrac{P \cos 45°}{\cos 20°} \sin 20°$

$19.6 = P\left(\sin 45° + \dfrac{\cos 45°}{\cos 20°} \sin 20\right)$

$P = 20.32\ldots$

Tip: You need to practise solving these types of simultaneous equations.

The tensions in strings are 20.3 N (3 s.f.) and 15.3 N (3 s.f.).

3.3 Friction and the coefficient of friction

Represent the contact force between two rough surfaces by two components, 'the normal force' and the 'frictional force', understand the concept of limiting friction and limiting equilibrium, recall the definition of coefficient of friction, and use the relationship $F \leqslant \mu \mathbf{R}$ or $F = \mu \mathbf{R}$ as appropriate.

When a particle of mass m kg rests on a rough surface, there is a frictional force **F** N which has a maximum value, F_{max} N. Suppose there is a pulling force of **P** N:

Consider the horizontal components of the forces; there are three situations that can arise:

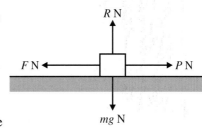

Recall: On a smooth surface, $\mathbf{F} = 0$ N.

Note: If there is no pulling force then the frictional force exerted = 0 N.

1 If the force $P < F_{max}$ then the frictional force $F = P$, i.e. friction will take a value sufficient to maintain equilibrium.

2 If the force $P = F_{max}$ then the situation is said to be in **limiting equilibrium** and the system is at the point of moving.

3 If the force $P > F_{max}$ then equilibrium is broken, the object will move and friction will maintain its maximum value, F_{max}.

When the situation is in limiting equilibrium then

$$F_{max} = \mu R$$

where μ is called the **coefficient of friction** and R is the normal reaction force of the particle. In this case friction is said to be **limiting**. μ is a property of surface; rough surfaces have large values of μ and exert a large force of friction. Similarly, large values of the normal reaction cause large forces of friction.

Note:
If the frictional force acting is F then:
$$0 < F \le F_{max}\ (= \mu \mathbf{R}).$$

Recall:
The normal reaction force is related to the weight.

Example 3.6 A car of mass 1000 kg is being pulled along a rough, horizontal surface by a rope. The tension in the rope is 5000 N and the rope is inclined at an angle of 15° above the horizontal. Find the normal reaction force and the least value of the coefficient of friction needed to maintain equilibrium.

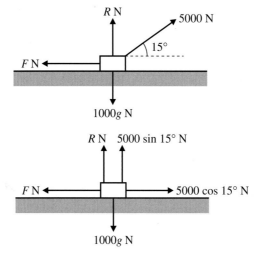

Step 1: Draw the force diagram and, where appropriate, redraw each of the forces resolved into any two perpendicular directions.

Note:
The method remains the same as in Sections 3.1 and 3.2; the only change is in the unknowns that you are asked to find.

Note:
Initially, assume the situation is in limiting equilibrium, so $\mathbf{F} = \mu \mathbf{R}$.
($\mathbf{F} = \mu \mathbf{R}$ only in limiting equilibrium.)

Horizontal components:

$5000 \cos 15° - F = 0$

$F = 5000 \cos 15°$

Limiting equilibrium: $F = \mu R$

Step 2: Find the resultant force in each direction (and equate each to 0 when in equilibrium).

Vertical components:

$R + 5000 \sin 15° - 1000g = 0$

$R = 1000g - 5000 \sin 15°$

$\qquad = 8505.9...$

The normal reaction is 8500 N (2 s.f.).

Note:
μ has no units.

Step 3: Solve for unknowns, using $F = \mu R$ when in limiting equilibrium.

$$\mu = \frac{5000 \cos 15°}{8505.9...}$$

The coefficient of friction is 0.57 (2 s.f.), when in limiting equilibrium.

Any higher value of μ means that equilibrium will be maintained. So $\mu \ge 0.57$, and the least value of the coefficient of friction needed to maintain equilibrium is 0.57 (2 s.f.).

Note:
Quote answers to 2 s.f. when using $g = 9.8$.

Note:
Another method of solving for μ is to use $\mathbf{F} \le \mu \mathbf{R}$
$\Rightarrow \mu \ge \dfrac{\mathbf{F}}{\mathbf{R}} = \dfrac{500 \cos 15°}{8505.9...}$
$\qquad \mu \ge 0.57$ (2 s.f.)

Forces on an inclined plane

If a particle is on an inclined plane then, instead of resolving horizontally and vertically, resolve parallel to the plane and perpendicular to the plane, as this simplifies the algebra.

Parallel and perpendicular components of the weight

If a particle of mass m kg rests on a plane inclined at angle θ to the horizontal, the weight acts vertically downwards:

Draw the force diagram:

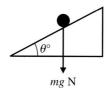

mg N

The components parallel and perpendicular to the plane are:

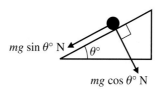

$mg \sin \theta°$ N

$mg \cos \theta°$ N

Note:
This can be seen by drawing a force triangle:

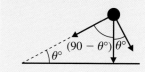

$(90 - \theta°)$ $\theta°$ $\theta°$

The parallel component of the weight is $mg \sin \theta$ down the plane.
The perpendicular component of the weight is $mg \cos \theta$.

Parallel and perpendicular components of the normal reaction force and friction

The normal reaction always acts perpendicular to the surface in contact so it only has a perpendicular component ($= R$). Similarly, the friction is always parallel to the plane and so it only has a parallel component, in a direction opposing relative motion.

Note:
There is no need to resolve the normal reaction force and friction.

Example 3.7 A particle of mass 4 kg rests on a smooth plane inclined at 30° to the horizontal. A force of P N acts on the particle up the plane along the line of greatest slope. Find the magnitude of the reaction force and the magnitude of the force P.

Step 1: Draw the force diagram and, where appropriate, redraw each of the forces resolved into any two perpendicular directions.

Draw the force diagram:

R N

P N

30°

$4g$ N

The components parallel and perpendicular to the plane are:

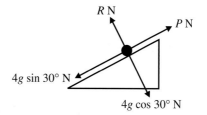

R N

P N

$4g \sin 30°$ N

$4g \cos 30°$ N

Note:
Resolve the forces in perpendicular directions (parallel and perpendicular to the plane).

Step 2: Find the resultant force in each direction (and equate each to 0 when in equilibrium).

Step 3: Solve for unknowns.

Perpendicular to the plane:
$$0 = R - 4g \cos 30°$$
$$R = 4g \cos 30°$$
$$= 33.9\ldots$$

Parallel to the plane:
$$0 = P - 4g \sin 30°$$
$$P = 4g \sin 30°$$
$$= 19.6$$

The reaction force has magnitude 34 N (2 s.f.) and the force P has magnitude 19.6 N.

Parallel and perpendicular components of a horizontal force

If a horizontal force **X** N acts on a particle on an inclined plane then:

- the component of **X** parallel to the plane is $X \cos \theta°$

- the component of **X** perpendicular to the plane is $X \sin \theta°$.

Draw the force diagram:

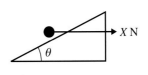

The parallel and perpendicular components are:

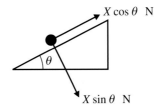

Note:

Note:
This can be seen by drawing a force triangle:

Example 3.8   A particle of mass 5 kg rests on a rough plane inclined at $\theta°$ to the horizontal where $\sin \theta° = 0.6$. The particle is kept in equilibrium by a horizontal force, of magnitude 40 N, acting in the vertical plane containing the line of greatest slope of the inclined plane through the particle. The particle is on the point of slipping up the plane.

a Find the force exerted by the plane on the particle.

b Find the coefficient of friction.

c How would the force diagram change if the particle were on the point of slipping down the plane?

Step 1: Draw the force diagram and, where appropriate, redraw each of the forces resolved into any two perpendicular directions.

a Draw the force diagram:

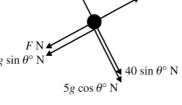

The components parallel and perpendicular to the plane are:

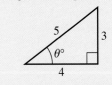

Note:
Friction acts to oppose the motion, down the plane.

Recall:
Pythagorean triangles.

$\cos \theta° = 0.8$, $\sin \theta° = 0.6$

Step 2: Find the resultant force in each direction (and equate each to 0 when in equilibrium).

Perpendicular to the plane:
$R - 40 \sin \theta° - 5g \cos \theta° = 0$
Substitute for $\cos \theta°$ and $\sin \theta°$:
$R - 24 - 4g = 0$
$R = 24 + 4g = 63.2$
The normal reaction is
63 N (2 s.f.).

Parallel to the plane:
$40 \cos \theta° - 5g \sin \theta° - F = 0$

$32 - 3g - F = 0$
$F = 2.6$

Tip:
Substitute known values for $\cos \theta°$ and $\sin \theta°$.

Step 3: Solve for unknowns, using $F = \mu R$ when in limiting equilibrium.

b In limiting equilibrium, $F = \mu R$.

$$\mu = \frac{F}{R} = \frac{2.6}{63.2} = 0.04113\ldots$$

Note:
The frictional force can be written as μR because the particle is in limiting equilibrium.

c If the particle was about to slip down the plane, the friction force would act up the plane, opposing the relative motion.

Parallel and perpendicular components of a force inclined at an angle $\alpha°$ to the plane

Draw the force diagram:

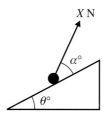

The components parallel and perpendicular to the plane are:

$X \sin \alpha°$ N

$X \cos \alpha°$ N

Note:
The force is being resolved relative to the inclined plane, so $\theta°$ is not involved in the parallel and perpendicular components.

The component of **X** parallel to the plane is $X \cos \alpha°$ (up the plane).

The component of **X** perpendicular to the plane is $X \sin \alpha°$.

Example 3.9   A child of mass 35 kg sits on a toboggan, which rests on a rough inclined plane of angle 45°. The toboggan is light. The child is being pulled upwards by a rope, which is inclined at an angle of 15° above the line of greatest slope of the plane. The coefficient of friction between the toboggan and the plane is 0.25. Equilibrium is about to be broken by the child sliding down the plane. By treating the child and the toboggan as a single particle, find the tension in the rope and the magnitude of the limiting frictional force.

Recall:
'The toboggan is light' means that you can ignore its weight when modelling the situation.

Step 1: Draw the force diagram and, where appropriate, redraw each of the forces resolved into any two perpendicular directions.

Draw the force diagram:

R N T N F N   15°   45° $35g$ N

The components parallel and perpendicular to the plane are:

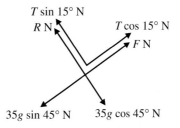

$T \sin 15°$ N R N $T \cos 15°$ N F N $35g \sin 45°$ N $35g \cos 45°$ N

Tip:
Let **T** be the tension in the rope.

Note:
Friction acts to oppose the motion, i.e. up the plane.

Step 2: Find the resultant force in each direction (and equate each to 0 when in equilibrium).

Step 3: Solve for unknowns, using $F = \mu R$ when in limiting equilibrium.

Parallel to the plane:

$T \cos 15° + F - 35g \sin 45° = 0$

$F = 35g \sin 45° - T \cos 15°$

Perpendicular to the plane:

$T \sin 15° + R - 35g \cos 45° = 0$

$R = 35g \cos 45° - T \sin 15°$

In limiting equilibrium: $F = \mu R$

$35g \sin 45° - T \cos 15° = 0.25 (35g \cos 45° - T \sin 15°)$

$T(\cos 15° - 0.25 \sin 15°) = 35g \sin 45° - 0.25(35g \cos 45°)$

$T = 201.84\ldots$

$R = 190.29\ldots$

$F = \mu R = 0.25R = 47.57\ldots$

The tension in the rope is 200 N and the limiting frictional force is 48 N (all 2 s.f.).

1 The diagram shows the forces **A** N, **B** N and **C** N acting on a particle. The forces are in equilibrium. Find **A** and **B** when:

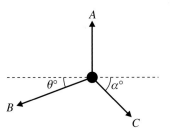

a $\theta = 25$, $\alpha = 40$, **C** = 10

b $\theta = 18$, $\alpha = 20$, **C** = 19.

Quote your answers to a suitable degree of accuracy.

 2 A particle is acted on by a force of 15 N which acts on a bearing of 020°, and another force of 4 N which acts on a bearing of 230°. Find the magnitude of a third force which will keep the system in equilibrium, stating the angle of its line of action as a bearing.

3 A particle of mass m kg rests on a rough horizontal plane. A horizontal force of P N acts on the particle. Find the magnitude of the normal reaction force and the frictional force if:

a $m = 0.2$, $P = 7$ **b** $m = 3$, $P = 14$ **c** $m = 9$, $P = 24$.

4 A particle of mass m kg lies at rest on a rough horizontal plane and is on the point of slipping. A string applies a tension T N at an angle 30° above the horizontal and pulls the particle along the plane. If the frictional force is F N along the plane and the normal reaction force of the plane on the particle is R N, find the magnitude of:

a F and R if $m = 5$ and $T = 12$

b T and m if $F = 12$ and $R = 2$

c T and R if $F = 6$ and $m = 0.9$.

 5 Two men try to push a car of mass 1500 kg, which lies on a rough horizontal road. The two men apply forces of 50 N and 100 N. The car does not move. No other forces are present in the horizontal plane except friction.

a Find the frictional force between the car and the road.

b Given that the car is about to move find the coefficient of friction between the car and the road.

6 A particle of mass m kg is at rest on a rough horizontal table. A string with tension T N acting at an angle of 15° below the horizontal pulls the particle along the plane. The coefficient of friction between the plane and the block is μ and the normal reaction force exerted by the plane on the block is R N. The situation is in equilibrium and friction is limiting; find the magnitude of:

a R and μ if $m = 2$ and $T = 20$

b T and R if $m = 0.5$ and $\mu = 0.25$

c μ and m if $T = 15$ and $R = 15$.

7 A particle of weight 49 N is suspended freely in the air by two light strings as shown in the diagram. Find the tensions in the two strings.

 8 The diagram shows a particle suspended from a horizontal beam by two unequal, light and inextensible strings. Given that the tension in the left string is 8 N and it makes an angle of 40° to the beam, and the other string makes an angle of 60° to the beam, find the tension in the other string and the mass of the particle.

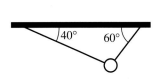

9 A particle of mass m kg lies on a rough plane, inclined at $\theta°$ to the horizontal where $\cos \theta° = \frac{12}{13}$. The system is in equilibrium. Find the magnitude of the normal reaction force and the frictional force, leaving your answer in terms of m and g, where g is the acceleration due to gravity.
Given also that the particle is in limiting equilibrium, find the coefficient of friction between the plane and the particle.

10 A particle of weight 29.4 N lies on a rough plane inclined at 30° to the horizontal. A force of X N acts up the plane, along the line of greatest slope of the plane. The coefficient of friction between the particle and the plane is 0.5. Find the magnitude of X if:

  a the particle is on the point of slipping up the plane

  b the particle is on the point of slipping down the plane.

11 **a** Repeat question **10a** with the force of X N parallel to the plane replaced by a horizontal force of X N acting in the vertical plane containing the line of greatest slope of the inclined plane through the particle.

  b Repeat question **10b** with the force of X N parallel to the plane replaced by a force of X N pulling the particle up the plane, acting at 45° above the line of greatest slope of the plane.

12 A car of mass 1500 kg is broken down on a rough plane inclined at an angle of $\theta°$ to the horizontal, where $\theta° = \sin^{-1}\left(\frac{7}{25}\right)$. It is being pulled up the plane by means of a towrope, which is acting at 9° above the line of greatest slope of the plane. The resistance between the plane and the car has magnitude 1000 N. The car is at rest in equilibrium and is about to move up the plane. Find the tension in the towrope and the magnitude of the force of the plane on the car.

SKILLS CHECK **3A EXTRA** is on the CD

Examination practice Equilibrium of a particle

1

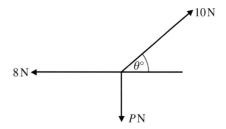

Three forces, of magnitudes 10 N, 8 N and P N, act at a point in the directions shown in the diagram. The forces are in equilibrium. Find

  i the value of θ,

  ii the value of P.

The force of magnitude 8 N is now removed.

  iii Find the magnitude and direction of the resultant of the two remaining forces.   [OCR Jan 2003]

2 A small block of mass 0.5 kg is acted on by a horizontal force of magnitude 1.96 N and is at rest on a horizontal plane.

  i Find the normal force exerted on the block by the plane.

  ii State the frictional force exerted on the block by the plane.

  iii Given that the block is on the point of slipping, find the coefficient of friction between the block and the plane.   [OCR Nov 2003]

3

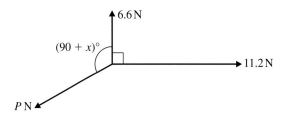

Three coplanar forces have magnitudes and directions as shown in the diagram.

i Given that the three forces are in equilibrium, find the values of P and x.

The force of magnitude P N is now removed.

ii Write down the magnitude and direction of the resultant of the two remaining forces.

[OCR Jan 2004]

4 A box of mass 200 kg rests in equilibrium on a plane inclined at 32° to the horizontal.

i Calculate the frictional force acting on the box.

ii Given that the equilibrium is limiting, calculate the coefficient of friction between the box and the plane.

iii Calculate the maximum force, acting up the slope, which can be applied to the box without causing the box to slip.

[OCR June 2003]

 5 i A crate of mass 250 kg rests in equilibrium on a slope inclined at 20° to the horizontal. Find the frictional force acting on the crate.

ii Given that the equilibrium is limiting, calculate the coefficient of friction between the crate and the slope.

iii The crate is now pushed horizontally with a force of magnitude 2000 N, as shown in the diagram below. Show that the crate remains in equilibrium.

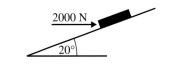

[OCR Jan 2001]

6

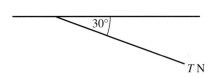

A heavy ring of mass 5 kg is threaded on a fixed rough horizontal rod. The coefficient of friction between the ring and the rod is $\frac{1}{2}$. A light string is attached to the ring and is pulled downwards with a force of magnitude T newtons acting at an angle of 30° to the horizontal (see diagram). Given that the ring is about to slip along the rod, find the value of T.

[OCR Specimen]

7

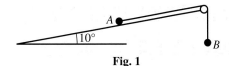

Fig. 1

Two particles A and B, of masses 0.5 kg and m kg respectively, are joined by a light inextensible string. Particle A rests on a board which is fixed at $10°$ to the horizontal. The coefficient of friction between A and the board is 0.15. The string is parallel to a line of greatest slope of the board and passes over a small smooth pulley at the top of the board. Particle B hangs vertically below the pulley, as shown in Fig. 1. The system is in equilibrium.

i **a** Given that A is on the point of sliding up the plane, show that $m = 0.161$, correct to 3 decimal places.

   b Given instead that A is on the point of sliding down the plane, show that $m = 0.013$, correct to 3 decimal places.

The board is now fixed horizontally and A is held on the board. The string passes over the pulley and B hangs vertically below the pulley as shown in Fig. 2. A is now released.

Fig. 2

ii Show that the system starts to move if m has the value in part **i a**, but remains at rest if m has the value in part **i b**.

iii Find the magnitude of the frictional force acting on A for each of the values of m in part **i**.

[OCR June 2001]

4 Kinematics of motion in a straight line

4.1 Distance, displacement, speed, velocity and acceleration

Understanding the concepts of distance and speed as scalar quantities, and of displacement, velocity and acceleration as vector quantities.

The following quantities are needed to describe the motion of a particle.

Distance is the length of a given path; it is a scalar quantity. We measure distance in metres, written m.

Displacement defines the position of one point relative to another point: displacement includes both the distance between the two points and the direction of the first point from the second. Displacement is measured in metres, written m. It is a vector quantity as it has both size and direction.

Speed is the rate of change of the distance with time, with no account taken for direction. The units for speed are metres per second, written m/s or $m\,s^{-1}$. Speed is a scalar quantity.

Velocity is the rate of change of displacement with time, and so the direction is taken into account. Thus it is a vector quantity. The units for velocity are metres per second, written m/s or $m\,s^{-1}$.

Acceleration is the rate of change of velocity with time. The units for acceleration are metres per second squared, written m/s^2 or $m\,s^{-2}$. Acceleration is a vector quantity.

4.2 Kinematics graphs

Sketch and interpret (t, x) and (t, v) graphs.

The aim of kinematics is to analyse the motion of a particle that is travelling in a straight line – its velocity, time of travel, acceleration and the path that it follows. In this section, the particle always has a constant acceleration.

The properties that you are analysing are (with standard units in brackets):
s = the displacement of the particle from a fixed point (metres, m)
u = the initial velocity of the particle (m/s or $m\,s^{-1}$)
v = the final velocity of the particle (m/s or $m\,s^{-1}$)
a = the acceleration of the particle (m/s^2 or $m\,s^{-2}$)
t = the time taken for the particle to travel (seconds, s).

A **speed–time graph** shows how the speed of a particle varies with time. The horizontal axis represents the time taken and the vertical axis represents the speed. In the graph, u is the initial velocity and v is the final velocity after travelling for t seconds under constant acceleration.

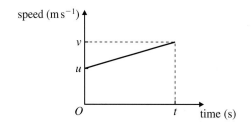

Recall:
Acceleration is the rate of change of velocity of a particle, where velocity is the speed of a particle in a given direction.

Note:
These variables are often referred to as *suvat*.

You can obtain important information from a speed–time graph.

The gradient of the line $= \dfrac{(v - u)}{t} = a$.

The area under the line $= \frac{1}{2}(u + v)t = s$.

Remember, in a speed–time graph:

the gradient of the line = the acceleration of the particle
the area under the line = the distance travelled by the particle

A **velocity–time** graph shows how the velocity of a particle varies with time. In this case the area represents the displacement, and *when the area is below the time axis this represents a negative displacement.*

A **displacement–time** graph shows how the displacement of a particle varies with time. The horizontal axis represents the time taken and the vertical axis represents the displacement. We also know that:

$$\text{velocity} = \frac{\text{change in displacement}}{\text{time taken}}, \text{ and so:}$$

the gradient of a displacement–time graph

= the velocity of the particle

A particle that slows down is **decelerating**. When this is the case, the acceleration is negative; this can also be called retardation.

Recall:
From C1, the gradient, m, between two points (x_1, y_1) and (x_2, y_2) is:
$$m = \frac{y_2 - y_1}{x_2 - x_1}$$

Recall:
The area of a trapezium is $\frac{1}{2}(a + b)h$, where a and b are the lengths of the parallel sides and h is the height.

Note:
Acceleration is the rate of change of velocity $= \dfrac{\text{change in velocity}}{\text{time}}$ when the acceleration is constant.

Note:
Velocity–time graphs can also be referred to as (t, v) graphs. Displacement–time graphs can also be referred to as (t, x) graphs.

Example 4.1   A particle starts from a point O and travels 50 m due north in 10 seconds to a point A. The particle then travels 70 m due south in 5 seconds to a point B.

    a Sketch a (t, x) graph of the journey of the particle from O to B.

    b Find the total change in displacement in its journey from O to B.

Step 1: Draw a clear diagram to represent the information given. **a**

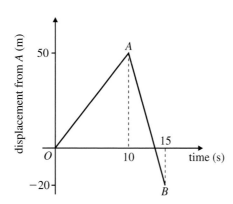

Step 2: Work out the displacement. **b** The total displacement from O to B is -20 m.

Example 4.2   A particle travels with constant speed 4 m s^{-1} from A to B for 4 seconds. It then turns around and travels in the opposite direction from B to A with constant speed 2 m s^{-1} for a further 2 seconds.

    a Sketch a (t, v) graph to represent the journey of the particle in the first 6 seconds.

    b Calculate the displacement after the first 4 seconds and after 6 seconds of the motion. Hence, sketch a displacement–time graph.

Note:
Distance = (average) speed × time.

Step 1: Draw a clear diagram to represent the information given.

a

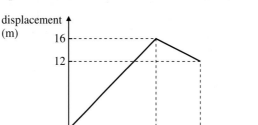

Step 2: Use area = displacement to solve the problem.

b Area under the graph for the first 4 seconds = $4 \times 4 = 16$.

So the displacement after the first 4 seconds is 16 m.

Area 'under' the graph for the last 2 seconds = $2 \times 2 = 4$.

So the displacement after 6 seconds is $16 + (-4) = 12$ m.

Recall:
The gradient must be constant for the displacement–time graph because the velocity is constant.

Example 4.3 A car starts from a point A and accelerates uniformly at 1 m s^{-2} from rest for three seconds. It then maintains a steady speed for a further 15 seconds. Finally it slows down for 5 seconds with uniform retardation until it stops at point B.

a Sketch a speed–time graph for the journey of the car.

b Find the maximum speed of the car.

c Find the acceleration of the car in the final stage of the journey.

d Find the distance AB.

Step 1: Draw a clear diagram to represent the information given.

a Let the maximum speed be $v \text{ m s}^{-1}$.

Step 2: Use gradient = acceleration and area = distance to solve the problem.

b Acceleration in first 3 seconds = gradient in first three seconds

$$1 = \frac{(v - 0)}{(3 - 0)}$$

$$v = 3$$

The maximum speed of the car is 3 m s^{-1}.

c Acceleration in last 5 seconds = gradient in last five seconds

$$a = \frac{0 - v}{5} = -\frac{3}{5}$$

The acceleration of the car in the last 5 seconds is -0.6 m s^{-2}.

Note:
You could also use the equations of motion, studied in Section 4.4, to solve this problem.

d Distance travelled from A to B = total area

$$\text{area of trapezium} = \tfrac{1}{2}(a + b)h$$
$$= \tfrac{1}{2} \times (23 + 15) \times 3$$
$$= 57$$

The distance AB is 57 m.

Note:
The area could also be calculated by dividing the trapezium into two triangles and a rectangle, and then summing their areas.

An **acceleration–time** graph shows how the acceleration of a particle varies with time. This can be sketched from a velocity–time graph by using the fact that the gradient of a velocity–time graph gives the acceleration.

Example 4.4 At $t = 0$, where t is the time in seconds, a train is travelling with a uniform velocity of 7 m s^{-1}. The train then approaches a station and, at $t = 4$, the driver applies the brake slowing down the train with uniform retardation. It reaches the station T seconds after applying the brakes, where it comes to a halt.

a Sketch a velocity–time graph of the journey of the train.

b Calculate the retardation, leaving your answer in terms of T.

The distance travelled by the train during the first 4 seconds is three quarters of the distance travelled during T seconds.

c Calculate the value of T and the retardation.

d Hence sketch an acceleration–time graph of the journey of the train.

Step 1: Draw a clear diagram to represent the information given.

a

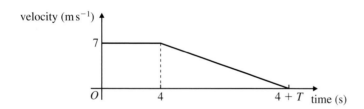

Step 2: Use gradient = acceleration and area = distance to solve the problem.

b Acceleration in last T seconds = gradient in last T seconds

$$a = \frac{(v - u)}{t} = \frac{(0 - 7)}{T} = -\frac{7}{T}$$

Hence, the retardation of the train in the last T seconds is $\dfrac{7}{T}$ m s^{-2}.

Recall:
Acceleration is given by the gradient in a speed–time graph.

Recall:
Distance travelled is given by the area under a speed–time graph.

c Distance travelled in first 4 s = 0.75 × (distance travelled in last T s)

$$4 \times 7 = 0.75\left(\frac{1}{2} \times T \times 7\right)$$
$$T = \frac{32}{3}$$

Note:
Substitute this value for T into the relation in **b**.

The train slows for $10\frac{2}{3}$ s.

$$\text{Retardation} = \frac{7}{T} = \frac{21}{32} = 0.65625$$

The retardation of the train is 0.656 m s^{-2} (3 s.f.).

d Acceleration–time graph:

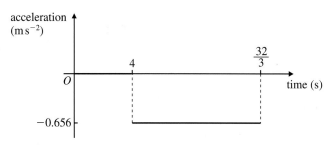

Example 4.5 A motorbike tries to catch up with a car. The car leaves 20 seconds before the motorbike, and travels with a constant speed of 8 m s^{-1}. The motorbike accelerates uniformly at 2 m s^{-2}, until it reaches a speed of 12 m s^{-1}. It then maintains this constant speed for the remainder of the journey. Find the time taken for the motorbike to reach the car.

Step 1: Draw a clear diagram to represent the information given.

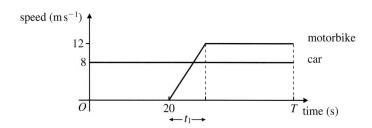

Note:
The car leaves at $t = 0$, the motorbike leaves at $t = 20$ and the two meet at $t = T$.

Step 2: Use gradient = acceleration and area = distance to solve the problem.

Acceleration of motorbike = gradient of speed–time graph during t_1 s.

$$2 = \frac{12}{t_1}$$

$$t_1 = 6$$

Tip:
First calculate the time for which the motorbike accelerates (t_1).

When the motorbike reaches the car

Area under line (car) = area under line (motorbike)

$$8 \times T = \tfrac{1}{2} \times 12 \times ((T - 20) + (T - 26))$$

$$T = 69$$

The motorbike catches up with the car 69 seconds after the car left.

Hence it takes the motorbike $(69 - 20)$ s $= 49$ s to reach the car.

Note:
When two particles meet, their distances, and hence the areas of their speed–time graphs, must be equal at the same given time.

4.3 Differentiation and integration of displacement, velocity and acceleration

Use differentiation and integration with respect to time to solve simple problems concerning displacement, velocity and acceleration.

Displacement, velocity and acceleration were introduced in Section 4.1. When the acceleration of a particle is not constant, you can use calculus to find the velocity and displacement of the particle.

If a particle is travelling in a straight line and the displacement, measured from a fixed point, is given as a function of time, or displacement = $s(t)$ m, then the velocity, v m s^{-1} is given by:

$$v = \text{rate of change of displacement with time}$$

$$= \frac{\mathrm{d}s}{\mathrm{d}t}$$

Note:
s is a function of t means that s is given as a formula in terms of t.

Recall:
From C1: if $y = x^n$, then $\frac{\mathrm{d}y}{\mathrm{d}x} = nx^{n-1}$.

and the acceleration, a m s^{-2}, is given by

$$a = \text{rate of change of velocity with time}$$

$$= \frac{dv}{dt}$$

$$= \frac{d^2s}{dt^2} \quad \left(\text{since } v = \frac{ds}{dt}\right)$$

In other words you differentiate $s(t)$ with respect to time to find $v(t)$ and differentiate $v(t)$ with respect to time to find $a(t)$.

Example 4.6 A particle P moves in a straight line such that the displacement s m from a fixed point O at time t seconds is given by $s = t^4 - 4t^2 + 1$. Find:

    a the initial displacement of P from O

    b the distance of P from O when $t = 1$

    c the velocity of P when $t = 1$

    d the acceleration of P when $t = 2$.

Step 1: Substitute t value(s) into s.

$$s = t^4 - 4t^2 + 1$$

a At $t = 0$, $s = (0)^4 - 4(0)^2 + 1 = 1$
The initial displacement of P is 1 m.

b At $t = 1$, $s = 1^4 - 4(1)^2 + 1 = -2$
The distance of P from O at $t = 1$ is 2 m.

Step 1: Differentiate $(s \rightarrow v \rightarrow a)$.

c Differentiating, $v = \dfrac{ds}{dt}$

$$= 4t^3 - 8t$$

Step 2: Substitute t value into v.

At $t = 1$, $v = 4(1)^3 - 8(1) = -4$
The velocity of P when $t = 1$ is -4 m s^{-1}.

Step 1: Differentiate $(s \rightarrow v \rightarrow a)$.

d Differentiating, $a = \dfrac{dv}{dt}$

$$= 12t^2 - 8$$

Step 2: Substitute t value into a.

At $t = 2$, $a = 12(2)^2 - 8 = 40$
The acceleration of P when $t = 1$ is 40 m s^{-2}.

> **Note:**
> The initial displacement refers to the displacement when $t = 0$.

> **Note:**
> The size of the displacement is the distance.

> **Note:**
> The speed, when $t = 1$ (= size of velocity) is 4 m s^{-1}.

Example 4.7 A particle moves in a straight line such that its velocity, measured in m s^{-1}, is given by $(6t^2 - 12t - 18)$, where t is the time measured in seconds. Find

    a the time when the particle is instantaneously at rest

    b the acceleration of the particle when the particle is instantaneously at rest

    c the time when the velocity of the particle is minimum

    d the minimum velocity of the particle.

a

$$v = 6t^2 - 12t - 18$$

When $v = 0$, $6t^2 - 12t - 18 = 0$

Dividing by 6, $t^2 - 2t - 3 = 0$

$$(t + 1)(t - 3) = 0$$

$t = -1$ or $t = 3$

The time when the particle is at rest is 3 seconds.

Note:
The particle is at rest is the same as saying $v = 0$.

Note:
$t = -1$ is not significant physically.

Step 1: Differentiate ($s \rightarrow v \rightarrow a$). **b**

$$v = 6t^2 - 12t - 18$$

Differentiating, $a = \dfrac{dv}{dt} = 12t - 12$

Step 2: Substitute *t* value into *a*.

At $t = 3$ (the time when the particle is at rest)

$$a = 12(3) - 12$$
$$= 24$$

The acceleration when the particle is at rest is 24 m s^{-2}.

Step 1: Solve for *t* value(s) (using *a* = 0). **c**

$$a = 12t - 12$$

When $a = 0$, $12t - 12 = 0$

$$t = 1$$

The minimum velocity is after 1 second.

Recall:
The time for the maximum/minimum velocity can be found by solving $\dfrac{dv}{dt} = 0$ or, in other words, solving $a = 0$.

Step 1: Substitute *t* value into *v*. **d**

$$v = 6t^2 - 12t - 18$$

At $t = 1$, $v = 6(1)^2 - 12(1) - 18 = -24$

The minimum velocity of the particle is -24 m s^{-1}.

If the acceleration is given as a function of time, then you can find the velocity and displacement by reversing the process of differentiation. This is done by **integration**.

If the acceleration is given as a function of time, or acceleration $= a(t)$ m s^{-2}, then the velocity, v m s^{-1}, is given by:

$$v = \int a(t)\, dt + c$$

and the displacement, s m, is given by:

$$s = \int v(t)\, dt + k$$

Note:
c and k are the constants of integration.

The constants of integration can be determined if more information is given in the question.

Example 4.8 A particle moves in a straight line so that its acceleration, a m s^{-2}, at time t seconds is given by $a = t - 3$. Initially, the particle is at rest and, at $t = 6$, its displacement is -20 m from a fixed point O on the line. Find:

a the velocity of the particle as a function of t, and hence the speed of the particle at $t = 2$

b the initial displacement of the particle from O.

Note:
It is useful to summarise the information given:
$a = t - 3$;
at $t = 0$, $v = 0$
at $t = 6$, $s = -20$.

Step 1: Integrate $(a \rightarrow v \rightarrow s)$, and find the constant of integration.

a
$$a = \frac{dv}{dt} = t - 3$$

Integrating, $\qquad v = \tfrac{1}{2}t^2 - 3t + c$

At $t = 0$, $v = 0$: $\qquad 0 = \tfrac{1}{2}(0)^2 - 3(0) + c$

$$c = 0$$

So $\qquad v = \tfrac{1}{2}t^2 - 3t$

Step 2: Substitute t value into v.

At $t = 2$, $v = \tfrac{1}{2}(2)^2 - 3(2) = -4$

The speed of the particle at $t = 2$ is $4\ \text{m s}^{-1}$.

Recall:
From C1
$$\int x^n dx = \frac{1}{n+1}x^{n+1} + c$$
$(n \neq -1)$.

Step 1: Integrate $(a \rightarrow v \rightarrow s)$, and find the constant of integration.

b
$$v = \frac{ds}{dt} = \tfrac{1}{2}t^2 - 3t$$

Integrating $\qquad s = \tfrac{1}{6}t^3 - \tfrac{3}{2}t^2 + k$

At $t = 6$, $s = -20$: $\; -20 = \tfrac{1}{6}(6)^3 - \tfrac{3}{2}(6)^2 + k$

$$= 36 - 54 + k$$

$$k = -2$$

So $\qquad s = \tfrac{1}{6}t^3 - \tfrac{3}{2}t^2 - 2$

Step 2: Substitute t value into s.

At $t = 0$, $s = -2$

The initial displacement of the particle from O is -2 m.

Note:
The distance of the particle from O ($=$ size of displacement) is 2 m.

Note:
To summarise:
differentiate

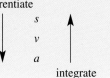

s

v

a

integrate

Example 4.9 A particle moves along the x-axis, so that at t seconds its velocity, $v\ \text{m s}^{-1}$, is given by $v = t - \tfrac{1}{2}t^2 - 4t^3$. The initial displacement of the particle from the origin, O, is $-\tfrac{1}{10}$ m. Find:

a the time when the particle reaches its maximum velocity, and hence the maximum velocity of the particle

b the displacement of the particle from O at the instant when the particle has its maximum velocity.

Step 1: Differentiate $(s \rightarrow v \rightarrow a)$.

a
$$v = t - \tfrac{1}{2}t^2 - 4t^3$$

Differentiating, $\qquad a = \frac{dv}{dt}$

$$= 1 - t - 12t^2$$

Recall:
The time for maximum/ minimum velocity can be found by solving $a = 0$.

Step 2: Solve for t value(s) (using $a = 0$).

$a = 0$, $\qquad 12t^2 + t - 1 = 0$

Factorising, $\quad (3t + 1)(4t - 1) = 0$

$t = -\tfrac{1}{3}$ or $t = \tfrac{1}{4}$

The time when the particle reaches its maximum velocity is $\tfrac{1}{4}$ seconds ($t = -\tfrac{1}{3}$ is not significant physically).

Step 3: Substitute t value into v.

At $t = \tfrac{1}{4}$, $v = \tfrac{1}{4} - \tfrac{1}{2}\left(\tfrac{1}{4}\right)^2 - 4\left(\tfrac{1}{4}\right)^3 = 0.15625$

The maximum velocity of the particle is $0.16\ \text{m s}^{-1}$ (2 s.f.).

Note:
It can be easily verified that this value of v is indeed a maximum, by substituting $t = \tfrac{1}{4}$ into the second derivative of v,
$\frac{d^2v}{dt^2}\left(= \frac{da}{dt}\right)$, which will be negative at $t = \tfrac{1}{4}$.

Step 1: Integrate
$(a \rightarrow v \rightarrow s)$, and find the
constant of integration.

b

$$v = \frac{ds}{dt} = t - \frac{1}{2}t^2 - 4t^3$$

Integrating, $s = \frac{1}{2}t^2 - \frac{1}{6}t^3 - t^4 + c$

At $t = 0, s = -\frac{1}{10}$, so $-\frac{1}{10} = c$

Hence, $s = \frac{1}{2}t^2 - \frac{1}{6}t^3 - t^4 - \frac{1}{10}$

Step 2: Substitute *t* value
into *s*.

At $t = \frac{1}{4}, s = \frac{1}{2}\left(\frac{1}{4}\right)^2 - \frac{1}{6}\left(\frac{1}{4}\right)^3 - \left(\frac{1}{4}\right)^4 - \frac{1}{10} = -0.075\ldots$

The displacement of the particle from O at the instant when the
particle has its maximum velocity is -0.075 m (2 s.f.).

SKILLS CHECK **4A: Kinematics graphs; differentiation and integration
of displacement, velocity and acceleration**

 1 A car leaves a point O and accelerates uniformly from rest to a speed of $4\,\text{m s}^{-1}$ in 2 seconds. It then
maintains a steady speed for a further 20 seconds, after which it slows down to a halt in 5 seconds.

   a Sketch a speed–time graph for the car's journey.

   b Find the acceleration during the initial and final stages of the journey.

   c Find the total distance travelled by the car.

2 A train travels between two stations, P and Q. The train accelerates uniformly from rest to $6\,\text{m s}^{-1}$ in
5 seconds. It then continues its journey for a further 15 seconds with this velocity, before
decelerating uniformly to a halt at station Q, in a further T seconds. The distance travelled in the last
T seconds is $\frac{1}{6}$ of the total distance travelled.

   a Sketch a velocity–time graph of the motion of the train between the two stations.

   b Find T.

   c Find the distance PQ.

   d By first finding the acceleration during each stage of the journey, sketch an acceleration–time
graph for the journey.

   e Find the average speed of the train's journey from P to Q.

 3 Two cars set off on a journey. The first car leaves at time $t = 0$, where t is measured in seconds. It
accelerates uniformly until it reaches a speed of $4\,\text{m s}^{-1}$ at $t = 3$. It then maintains a constant
velocity. The second car leaves from the same point at $t = 3$ and travels with constant speed $8\,\text{m s}^{-1}$.

   a On the same axes sketch speed–time graphs for the motion of the two cars.

   b Find t when the two cars meet.

   c How far are they from the start at this time?

   d Sketch a displacement–time graph for the journey of the second car.

4 A particle moves in a straight line. Its displacement, s m, from a fixed point O, after t seconds on that
line is given by $s = 2t^3 + 4t - 3$. Find:

   a the initial displacement of the particle from O

   b the velocity of the particle when $t = 1$

   c the acceleration of the particle when $t = 2$.

5 A particle P moves in a straight line so that at t seconds the displacement, s m, from a fixed point O is given by $s = t^3 - \dfrac{7}{2}t^2 + 2t - 4$. Find:

   a the velocity of P in terms of t

   b the values of the times, t_1 and t_2 seconds (where $0 < t_1 < t_2$), when the particle is instantaneously at rest

   c the acceleration of P at t_1 and t_2 seconds.

6 A particle moves in a straight line, so that at t seconds its acceleration, a m s^{-2}, is given by $a = 6t - 10$. Initially, the particle is at a fixed point O and has velocity 4 m s^{-1}. Find:

   a the velocity of the particle as a function of t

   b the speed of the particle at $t = 2$

   c the displacement of the particle from O as a function of t

   d the displacement of the particle from O when $t = 2$

   e the distance of the particle from O when $t = 3$

   f the times when the particle is again at O.

 7 A particle Q moves in a straight line so that, at t seconds after leaving a fixed point O, its acceleration is $-\dfrac{t}{5}$ m s^{-2}. At $t = 0$, the velocity of the particle is U m s^{-1}.

   a Find an expression for the velocity of Q in terms of t and U. Hence, find the value of U, given that the particle is instantaneously at rest at $t = 20$.

   b Find an expression for the displacement, s m, from O at t seconds.

   c Find the speed with which the particle returns to O.

8 A particle P is moving along the x-axis with velocity v m s^{-1}, given by $v = 4t - 2t^2$, where t is the time measured in seconds. When $t = 3$, the displacement of the particle from the origin, O, is 2 m. Find:

   a the position of P when $t = 2$

   b the acceleration of P in terms of t

   c the maximum velocity attained by P

   d the distance OP when the velocity is maximum.

9 A particle moves along the x-axis, passing through the origin O with speed 4 m s^{-1} in the positive x-direction. At time t seconds after passing through O the acceleration of the particle is $(2t - 5)$ m s^{-2}.

   a Find the value of t when the velocity is again 4 m s^{-1}.

   b Find the times when the particle is instantaneously at rest.

   c Find the displacement of the particle from O at time t seconds.

   d Hence, find the distance travelled by the particle in the third second, i.e. from $t = 2$ to $t = 3$.

10 A particle P moves in a straight line so that its displacement, s m, t seconds after leaving a fixed point O is given by $s = t^3 - 12t^2 - 144t$. Calculate:

   a the distance travelled by P between the instants when $t = 3$ and $t = 4$

   b calculate the time at which P returns to O, giving your answer to one decimal place

   c the value of s when P comes instantaneously to rest

   d the time when the acceleration of P is zero

   e the minimum velocity of P, verifying that the velocity you have found is indeed a minimum.

SKILLS CHECK **4A EXTRA** is on the CD

When the acceleration is constant or uniform, then the variables s, u, v, a and t (referred to as *suvat*) can be related using equations.

These **equations of motion** are called the **constant acceleration equations.** They are:

$v = u + at$

$s = ut + \frac{1}{2}at^2$

$s = \frac{1}{2}(u + v)t$

$v^2 = u^2 + 2as$

$s = vt - \frac{1}{2}at^2$

Note:
The units must be consistent when using these equations.

Each equation contains four variables. In most questions you will be given three variables and asked to calculate the fourth.

Example 4.10　A particle moves in a straight line from A to B. The particle starts from rest at A and accelerates at $2 \, \text{m s}^{-2}$ until it reaches a speed of $8 \, \text{m s}^{-1}$ at B.

  a　Find how long it takes to travel from A to B.

  b　Find the distance AB.

Tip:
If a particle starts from rest, this is another way of saying $u = 0$.

Step 1: Draw a clear diagram to represent the information given.

$$0 \, \text{m s}^{-1} \qquad\qquad \overset{2 \, \text{m s}^{-2}}{\longrightarrow} \qquad\qquad 8 \, \text{m s}^{-1}$$
$$A \rule{6cm}{0.4pt} B$$

Note:
Acceleration is represented by a double arrow $\longrightarrow\!\!\!\!\rightarrow$ and velocity by a single arrow $\longrightarrow$.

Step 2: Fill the information in *suvat*, identifying what is required with a question mark.

a　s　not required

$u = 0$

$v = 8$

$a = 2$

$t = ?$

b　$s = ?$

$u = 0$

$v = 8$

$a = 2$

t　not required

Step 3: Pick an equation of motion relating the three known variables with the unknown that is required, insert values, rearrange (if necessary) and solve.

$v = u + at$

$8 = 0 + 2t$

$t = 4$

It takes 4 s to travel from A to B.

$v^2 = u^2 + 2as$

$8^2 = 0^2 + 2(2)s$

$s = 16$

Distance AB is 16 m.

Tip:
Don't forget the units.

A particle that slows down is **decelerating**, in which case the acceleration is negative. If the acceleration of a particle is $-7 \, \text{m s}^{-2}$, then its **deceleration** or **retardation** is $7 \, \text{m s}^{-2}$.

Example 4.11　A train is travelling at $60 \, \text{km h}^{-1}$ on a straight railway track. The driver of the train sees a red signal 500 m ahead. He immediately applies the brakes so that the train decelerates at $0.25 \, \text{m s}^{-2}$.

  a　Find how far the train is past the signal when it comes to a halt.

  b　What is the velocity of the train as it passes the signal?

Note:
You must work in consistent units.
$1000 \, \text{m} = 1 \, \text{km}$ and $3600 \, \text{s} = 1$ hour, so to convert from km h^{-1} to m s^{-1} you multiply by 1000 and divide by 3600. In this case,
$60 \, \text{km h}^{-1} = 16\frac{2}{3} \, \text{m s}^{-1}$

Step 1: Draw a clear diagram to represent the information given.

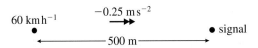

Step 2: Fill the information in *suvat*, identifying what is required with a question mark.

a Motion of train from the start until it stops:

$s = ?$
$u = 16\frac{2}{3}$
$v = 0$
$a = -0.25$
t not required

$v^2 = u^2 + 2as$
$(0)^2 = (16\frac{2}{3})^2 + 2(-0.25)s$
$s = 556$ (3 s.f.)

The train stops $(556 - 500)$ m $= 56$ m after the signal.

b Motion of train from the start until it passes the signal:

$s = 500$
$u = 16\frac{2}{3}$
$v = ?$
$a = -0.25$
t not required

$v^2 = u^2 + 2as$
$v^2 = (16\frac{2}{3})^2 + 2(-0.25)500$
$v = 5.27$ (3 s.f.)

The train travels at 5.27 m s^{-1} as it passes the red signal.

Step 3: Pick an equation of motion relating the three known variables with the unknown that is required, insert values, rearrange (if necessary) and solve.

> **Note:**
> Be careful not to confuse the variable s (displacement) with the unit s (seconds).

> **Note:**
> Only the positive root is physically reasonable.

Example 4.12   Two cars, *A* and *B*, start from the same point *O* and travel in a straight line. Car *A* leaves at $t = 0$ seconds and travels with constant speed 5 m s^{-1}. Car *B* leaves 10 seconds later and accelerates uniformly at 5 m s^{-2}. Find the distance from *O* when car *B* overtakes car *A*.

Suppose they meet *T* seconds after car *A* leaves *O*. t_A and t_B are the times for each car, measured from the start of *A*'s journey:

> **Note:**
> A particle travels with constant speed is another way of saying $a = 0$.

> **Note:**
> Consider the journey to the point where both cars have travelled the same distance.

Step 1: Draw a clear diagram to represent the information given.

For *A*: $t_A = 0$, $u = 5$ ———— $a = 0$ ———— $t_A = T$ (from O)

For *B*: $t_B = 10$, $u = 0$ ———— $a = 5$ ———— $t_B = T$ (from O)

Step 2: Fill the information in *suvat*, identifying what is required with a question mark.

For car *A*:
$s = ?$
$u = 5$
v not required
$a = 0$
$t = T$

For car *B*:
$s = ?$
$u = 0$
v not required
$a = 5$
$t = T - 10$

> **Note:**
> When there are two motions, apply *suvat* to each situation and find the common link. In this case, if car *A* has travelled for *T* s then car *B* has travelled for $(T - 10)$ s.

Step 3: Pick an equation of motion relating the three known variables with the unknown that is required, insert values, rearrange (if necessary) and solve.

$s = ut + \frac{1}{2}at^2$
$s = 5T + \frac{1}{2}(0)T^2$
$= 5T$

$s = ut + \frac{1}{2}at^2$
$s = (0)T + \frac{1}{2}(5)(T-10)^2$
$= 2.5T^2 - 50T + 250$

> **Note:**
> Pick equations that include both *s* and *t*.

When *B* overtakes *A*
$5T = 2.5T^2 - 50T + 250$
$0 = 2.5T^2 - 55T + 250$
$T = 15.58...$ or $6.42...$

> **Recall:**
> Using the quadratic formula from C1.

$T = 15.58...$ is the time that they meet because $T = 6.42...$ would give a negative time *t* for car *B*.
The cars have travelled $5 \times 15.58... = 77.9$ m (3 s.f.) when they meet.

Example 4.13   A particle moves in a straight line. It passes a point *O* with velocity u m s^{-1} and has constant acceleration a m s^{-2}. Two seconds later it passes a point *P*. One second after it passes the point *P* it passes a point *Q*. Given that the distance *OP* is 34 m and the distance *PQ* is 20 m, find *u* and *a*.

Step 1: Draw a clear diagram to represent the information given.

←——34 m——→ ←—20 m—→
•   •   •
O   *P*   *Q*
$t = 0$ $t = 2$ $t = 3$

33

Step 2: Fill the information in *suvat*, identifying what is required with a question mark.

Motion from O to P:

$s = 34$

$u = ?$

v not required

$a = ?$

$t = 2$

Motion from O to Q:

$s = 54$

$u = ?$

v not required

$a = ?$

$t = 3$

Note:
Analyse the motion from O to P and from O to Q because both sets of equations will have the same two unknowns, u and a.

Step 3: Pick an equation of motion relating the three known variables with the unknown that is required, insert values, rearrange (if necessary) and solve.

$s = ut + \frac{1}{2}at^2$

$34 = 2u + 2a$

$17 = u + a$ ①

$s = ut + \frac{1}{2}at^2$

$54 = 3u + 4.5a$

$18 = u + 1.5a$ ②

Note:
In this case there are two equations and two unknowns: solve them simultaneously.

Solving ① and ② simultaneously gives $u = 15$ and $a = 2$.

SKILLS CHECK **4B: Constant acceleration equations**

1 A particle travels in a straight horizontal line. It accelerates uniformly from rest to a speed of $4 \, \text{m s}^{-1}$ in 2 seconds.

   a Find the acceleration.

   b Find the distance travelled by the particle.

2 An object undergoing uniform acceleration travels 100 m in 12 seconds. Given the initial velocity is zero, find the acceleration.

3 A particle moves along a straight line AB with constant acceleration of $1 \, \text{m s}^{-2}$. If AB is 20 m and it takes 2 seconds to travel from A to B, what was the velocity of the particle at A?

4 A car travels in a straight line with uniform acceleration of $3 \, \text{m s}^{-2}$.

   a If the initial velocity is $8 \, \text{m s}^{-1}$, how long will it take to travel 6 m?

   b What assumptions have you made in modelling this situation?

5 A man drives a car with a constant acceleration of $5 \, \text{m s}^{-2}$. After 2 seconds he sees a set of traffic lights and slows down with a retardation of $2 \, \text{m s}^{-2}$. Given that his initial velocity was $2 \, \text{m s}^{-1}$ and that he manages to stop at the traffic lights, find the distance between the traffic lights and the point where he begins to accelerate.

6 A jet starts from rest and travels in a straight line with a constant acceleration of $10 \, \text{m s}^{-2}$.

   a How long will it take to reach a speed of $500 \, \text{km h}^{-1}$?

   b How far will it travel in this time?

 7 A particle moves in a straight line with constant acceleration. At $t = 0$, $t = 4$ and $t = 8$, where t is the time measured in seconds, the particle passes points P, Q and R respectively. The distance PQ is 100 m and the velocity at P is $20 \, \text{m s}^{-1}$.

   a Find the acceleration.

   b Find the distance QR.

8 A cyclist leaves a point O from rest and accelerates at $1 \, \text{m s}^{-2}$. Three seconds later a motorist, wishing to catch the cyclist, leaves from rest from the same point O and accelerates at $8 \, \text{m s}^{-2}$.

   a How far apart are the car and the bicycle 4 seconds after the cyclist left O?

   b After how long do they meet?

   c How far are they from O when they meet?

SKILLS CHECK **4B EXTRA is on the CD**

Motion in a vertical plane

Any particle travelling vertically experiences a constant acceleration towards the Earth due to gravity of magnitude $9.8\,\text{m s}^{-2}$. Gravity can be assumed to be constant (this is a fair modelling assumption), so the equations of motion can be applied.

Recall:
Modelling assumptions in Chapter 1.

Taking downwards as positive, the particle shown has a velocity of $-6\,\text{m s}^{-1}$ and an acceleration of $9.8\,\text{m s}^{-2}$.

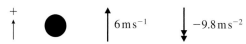

Note:
You can define either up or down as positive but be consistent within a problem.

Taking upwards as positive, the particle shown has a velocity of $6\,\text{m s}^{-1}$ and an acceleration of $-9.8\,\text{m s}^{-2}$.

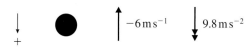

Example 4.14 A ball is thrown upwards with speed $18\,\text{m s}^{-1}$.

    a Find the displacement of the particle after 1 second.

    b Find the velocity after **i** 1 second and **ii** 2 seconds.

Recall:
Displacement is always measured from the starting point.

Step 1: Draw a clear diagram to represent the information given.

Taking upwards as positive:

Note:
Decide at the start which direction (up or down) is positive. In this case, up is positive.

Step 2: Fill the information in *suvat*, identifying what is required with a question mark.

a $s = ?$
     $u = 18$
     v not required
     $a = -9.8$
     $t = 1$

b s not required
     $u = 18$
     $v = ?$
     $a = -9.8$
     $t = 1$ and 2

Step 3: Pick an equation of motion relating the three known variables with the unknown that is required, insert values, rearrange (if necessary) and solve.

$s = ut + \frac{1}{2}at^2$
$s = (18 \times 1) + \frac{1}{2}(-9.8)\times 1^2$
   $= 13.1$

After 1 second the particle has a displacement of $13\,\text{m}$ (2 s.f.).

$v = u + at$
When $t = 1$
$v = 18 - (9.8 \times 1)$
   $= 8.2$

i   After 1 second the particle has a velocity of $8.2\,\text{m s}^{-1}$.

ii   When $t = 2$
     $v = 18 - (9.8 \times 2) = -1.6$
     After 2 seconds the particle has a velocity of $-1.6\,\text{m s}^{-1}$.

Note:
The negative sign indicates that the particle is travelling downwards.

Example 4.15 A particle is thrown downwards in the air and moves freely under gravity. It reaches twice its initial velocity after 2 seconds.

    a Find the initial velocity of the particle.

    b Find the displacement of the particle after 2 seconds.

Taking down as positive:

Note:
In this example, it is better to define down as positive because there is no motion upwards.

Step 1: Draw a clear diagram to represent the information given.

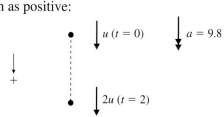

Step 2: Fill the information in *suvat*, identifying what is required with a question mark.

a
s not required
$u = ?$
$v = 2u$
$a = 9.8$
$t = 2$

b
$s = ?$
$u = 19.6$
$v = 39.2$
$a = 9.8$
$t = 2$

Step 3: Pick an equation of motion relating the three known variables with the unknown that is required, insert values, rearrange (if necessary) and solve.

$$v = u + at$$

$$2u = u + 9.8 \times 2$$

$$u = 19.6$$

The initial velocity of the particle is 20 m s^{-1} (2 s.f.).

$$s = \frac{(u + v)}{2}t$$

$$s = \frac{(19.6 + 39.2)}{2} \times 2$$

$$= 58.8$$

The displacement of the particle after 2 seconds is 59 m (2 s.f.).

The **maximum height** of a particle when it is thrown upwards is the height at which it stops travelling up and starts falling down. At this height its velocity in the vertical direction is 0 m s^{-1}. This is an important condition when calculating maximum height reached.

Example 4.16 A cricket ball is thrown upwards at a speed of 14 m s^{-1}. By modelling the ball as a particle, find **a** the maximum height reached and **b** the total distance travelled when it has come back to its starting point.

Taking up as positive:

Step 1: Draw a clear diagram to represent the information given.

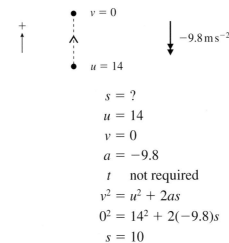

Step 2: Fill the information in *suvat*, identifying what is required with a question mark.

$s = ?$
$u = 14$
$v = 0$
$a = -9.8$
t not required

Note:
Using $v = 0$ at the maximum height defines the third piece of information required to fill *suvat*.

Step 3: Pick an equation of motion relating the three known variables with the unknown that is required, insert values, rearrange (if necessary) and solve.

$$v^2 = u^2 + 2as$$

$$0^2 = 14^2 + 2(-9.8)s$$

$$s = 10$$

a The maximum height reached is 10 m.

b The total distance travelled is $2 \times 10 = 20$ m.

Note:
From the start to the end of the motion the displacement is 0 m. Displacement is measured from the starting point.

The **time of flight** is the total time that a particle is in the air.

Example 4.17 A girl throws a ball vertically upwards in the air with speed 15 m s^{-1} and the ball travels freely under gravity. Find:

a the time of flight of the ball, assuming that it is thrown from the ground

b the time for which the ball is above a height of 2 m.

Step 1: Draw a clear diagram to represent the information given.

Defining up as positive:

$+$
$\uparrow$
$\bullet \; u = 15$
$\downarrow \; a = -9.8$

Step 2: Fill the information in *suvat*, identifying what is required with a question mark.

a
$s = 0$
$u = 15$
v not required
$a = -9.8$
$t = ?$

b
$s = 2$
$u = 15$
v not required
$a = -9.8$
$t = ?$

Step 3: Pick an equation of motion relating the three known variables with the unknown that is required, insert values, rearrange (if necessary) and solve.

a
$s = ut + \frac{1}{2}at^2$
$0 = 15t - 4.9t^2$
$0 = t(15 - 4.9t)$
$t = 0$ or $3.06\ldots$
The time of flight is 3.1 s (2 s.f.).

b
$s = ut + \frac{1}{2}at^2$
$2 = 15t - 4.9t^2$
$0 = 4.9t^2 - 15t + 2$
$t = 0.140\ldots$ or $2.92\ldots$
The ball is above 2 m for $2.9 - 0.14 = 2.8$ s (2 s.f.).

Tip:
These are the two times for which the particle is at the given displacement.

Example 4.18   A man throws a ball vertically upwards with a speed of 15 m s^{-1} from a height of 1.5 m above the ground. The ball travels freely under gravity.

  a   Calculate the time of flight.

  b   Calculate the speed with which the ball hits the ground.

Step 1: Draw a clear diagram to represent the information given.

Defining up as positive:

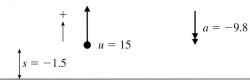

Step 2: Fill the information in *suvat*, identifying what is required with a question mark.

a
$s = -1.5$
$u = 15$
v not required
$a = -9.8$
$t = ?$

b
$s = -1.5$
$u = 15$
$v = ?$
$a = -9.8$
t not required

Note:
$s = 0$ at the height from which the ball is thrown and $s = -1.5$ at the ground.

Step 3: Pick an equation of motion relating the three known variables with the unknown that is required, insert values, rearrange (if necessary) and solve.

a
$s = ut + \frac{1}{2}at^2$
$-1.5 = 15t - 4.9t^2$
$4.9t^2 - 15t - 1.5 = 0$
$t = 3.15\ldots$ or $-0.096\ldots$
The time of flight is 3.2 s (2 s.f.).

b
$v^2 = u^2 + 2as$
$v^2 = 15^2 + 2(-9.8)(-1.5)$
$v = -15.9$
The speed with which the ball hits the ground is 16 m s^{-1} (2 s.f.).

Note:
The similarity between Example 4.17**a** and Example 4.18**a**.

SKILLS CHECK   **4C: Motion in a vertical plane**

1   A ball is thrown vertically upwards with velocity 28 m s^{-1} and travels freely under gravity. Find the velocity after two seconds and the distance that the particle has travelled from the start at this time.

2   Find the maximum height reached and the time of flight (the time taken to reach the start again) for the ball in question **1**. Also, find for how long it is above a height of 20 m.

3   A book is dropped from 80 m above ground and travels freely under gravity. Find the time taken for it to reach the ground. What assumptions have you made when modelling this situation?

 4   A stone is thrown vertically upwards from 4 m above ground. The initial velocity of the stone is 14 m s^{-1}. Find:

  a   the maximum height above the ground reached by the stone

  b   the time taken for it to hit the ground.

5   A ball is thrown upwards and reaches a height of 50 m. Find its initial velocity.

6 A ball is dropped from a window at $t = 0$, where t is the time in seconds. At $t = 2$ another ball is thrown from the same point with downwards velocity 25 m s^{-1}. Given that the balls travel freely under gravity, find:

    a the time when the balls pass each other

    b the distance from the window when they pass each other.

7 a Show that $h = \dfrac{u^2}{2g}$ for a particle travelling freely under gravity in the vertical plane with an initial velocity of u m s^{-1} (upwards) and a maximum height of h m above its starting point, where g is the acceleration due to gravity.

    b Calculate the difference in the maximum heights of two particles, travelling freely under gravity, given that they are thrown upwards with velocities 10 and 15 m s^{-1} respectively.

SKILLS CHECK 4C EXTRA is on the CD

Examination practice   Kinematics of motion in a straight line

1 A particle travels in a straight line. Its displacement from the fixed point O of the line, at time t seconds after leaving O, is $5t^2 - \frac{1}{2}t^3$ metres.

    i Write down expressions, in terms of t, for the velocity and acceleration of the particle.

    ii Find the velocity of the particle when its acceleration is 4 m s^{-2}.   [OCR June 2003]

2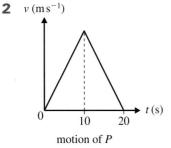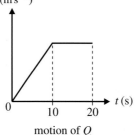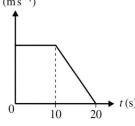

    motion of P motion of Q motion of R

    The three (t, v) graphs in the diagram refer to three particles P, Q and R, each of which travels in a straight line for 20 s. State which of P, Q and R

    i does **not** start from rest,

    ii is **not** slowing down for $10 < t < 20$.

    The acceleration of P, for $0 < t < 10$, is 0.3 m s^{-2}.

    iii Find the greatest speed of P.

    iv Find the distance travelled by P.

    v Given that Q travels the same distance as P, find the greatest speed of Q.   [OCR Jan 2003]

3 A particle leaves a point A with speed 1 m s^{-1} and travels with constant acceleration in a straight line to a point B, taking 50 s. The distance AB is 200 m.

    i Find the acceleration of the particle.

    ii Find the speed of the particle as it passes through the mid-point of AB.   [OCR Jan 2004]

4 A particle P moves in a straight line so that, at time t seconds after leaving a fixed point O, its acceleration is $-\frac{1}{10}t$ m s^{-2}. At time $t = 0$, the velocity of P is V m s^{-1}.

   i Find, by integration, an expression in terms of t and V for the velocity of P.

   ii Find the value of V, given that P is instantaneously at rest when $t = 10$.

   iii Find the displacement of P from O when $t = 10$.　　　　　　　[OCR June 2002]

5 A particle is projected vertically upwards from a fixed point O. The speed of projection is 5.6 m s^{-1}. At time T seconds after projection the particle is at a height of 0.7 m above O.

   i Find the two possible values of T, stating in each case whether the particle is travelling upwards or downwards.

   ii Find the total time for which the particle is more than 0.7 m above O.

   iii Find the height of the particle, above O, when its speed is 2.8 m s^{-1}.　　[OCR Jan 2003]

 6 Cyclists A and B are moving alongside each other in parallel straight lines, each at a constant speed of 7.5 m s^{-1}, when A starts to decelerate to pick up a bottle of water. His speed as he picks up the bottle is 1.5 m s^{-1}. Cyclist A then accelerates until he reaches his original speed of 7.5 m s^{-1}. Cyclist B continues at 7.5 m s^{-1} throughout.

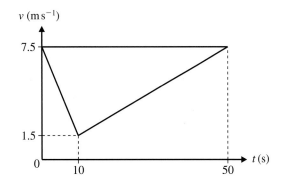

The diagram shows the (t, v) graph for A's motion from the instant he starts to decelerate until he regains his original speed. The graph consists of two straight line segments. Find:

   i the acceleration of A during the 40 s after picking up the bottle,

   ii the distance between B and A at the instant when A regains his initial speed,

   iii the time interval between B's arrival and A's arrival at the point where A regains his initial speed.　　　　　　　　　　　　　　　　[OCR Jan 2004]

7
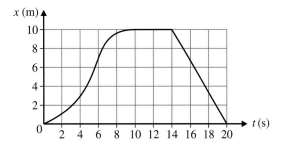

The diagram shows the (t, x) graph for the motion of a particle during the interval $0 \leqslant t \leqslant 20$, where x is in metres and t is in seconds.

   i Use the graph to find the maximum speed of the particle while it is moving away from its starting point. Give your answer in m s^{-1} to 1 significant figure.

   ii State how long the particle is at rest during the interval $0 \leqslant t \leqslant 20$.

   iii Find the speed of the particle while it is moving towards its starting point.　[OCR June 2004]

8 A particle is projected vertically upwards from ground level, with speed u m s^{-1}. The particle returns to the ground 6 s later. Find:

   i the value of u,

   ii the maximum height reached by the particle.

   iii the time for which the particle is at a height greater than half of its maximum height. [OCR Nov 2003]

 9 A particle moves in a straight line. Its velocity at time t s is v m s^{-1}, where $v = 0.3t^2 + 4t$.

   i Find an expression for the acceleration of the particle in terms of t.

The particle passes through the point A with velocity 20.8 m s^{-1}.

   ii Find the displacement of the particle from A when its acceleration is 7 m s^{-2}. [OCR Nov 2003]

 10 A particle is projected vertically upwards, from the ground, with a speed of 28 m s^{-1}. Ignoring air resistance, find

   i the maximum height reached by the particle,

   ii the speed of the particle when it is 30 m above the ground,

   iii the time taken for the particle to fall from its highest point to a height of 30 m,

   iv the length of time for which the particle is more than 30 m above the ground. [OCR June 2002]

5 Newton's laws of motion

5.1 Newton's laws of motion

Apply Newton's laws of motion to the linear motion of bodies of constant mass moving under the action of constant forces.

Definition of a force

A force acting on an object causes the object to accelerate. The unit of force is the **newton** (N). A force of 1 N acting on a particle of mass 1 kg causes it to accelerate at 1 m s^{-2}.

Newton's first law states that a particle will remain at rest or will continue to move with constant speed in one direction unless acted on by an external (resultant) force. In other words, a change in velocity of an object is caused by the action of a resultant force on the particle, otherwise it remains at rest or maintains constant velocity.

Newton's second law states that the resultant force, F N, produces an acceleration, a, that is proportional to the resultant force according to:

$$\mathbf{F} = ma$$

where m is the mass of the particle.

Note:
Learn this equation; remember $\mathbf{F}$ is the resultant force.

Newton's third law states that every action has an equal and opposite reaction, i.e. if one particle applies a force on another particle, the other particle applies an equal force on the first but in the opposite direction. This is the principle behind the normal reaction force, $\mathbf{R}$ (the surface exerts an equal and opposite force on the particle that rests on the surface).

Recall:
The normal reaction $\mathbf{R}$ is equal to the weight of a particle, mg, on a horizontal surface when no other forces act with a vertical component.

$\mathbf{F} = ma$ describes motion in all possible directions. The particle will accelerate in the direction of the resultant force.

Horizontal motion

When the force acts horizontally on an object it will accelerate in the horizontal direction according to $\mathbf{F} = ma$.

Example 5.1 A particle of mass 3 kg lies on a smooth horizontal plane. A horizontal force of 18 N acts on the particle. Calculate the acceleration of the particle.

Step 1: Draw the force diagram resolving the forces into any two perpendicular directions.

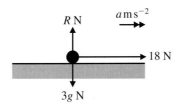

Note:
You do not need to resolve vertically in this example.

Step 2: Find the resultant force in each direction (and equate each to 0 if in equilibrium or to ma if accelerating).

Step 3: Solve for unknowns.

Horizontal components:

$F_h = ma$
$18 = 3a$
$a = 6$

Vertical components:

The particle is in equilibrium.

The particle accelerates at 6 m s^{-2} in the direction of the 18 N force.

Example 5.2   A particle of mass 5 kg is being pulled along a rough horizontal plane by a horizontal force of magnitude 15 N against a constant frictional force of magnitude 10 N. Given that the particle is initially at rest find:

a   the acceleration of the particle

b   the distance travelled by the particle in the first 3 seconds.

Step 1: Draw the force diagram (resolving the forces into any two perpendicular directions where necessary).

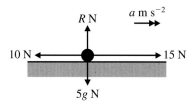

Step 2: Find the resultant force in each direction (and equate each to 0 if in equilibrium or to *ma* if accelerating).

Step 3: Solve for unknowns.

a   Horizontal components:

$$F_h = ma$$
$$15 - 10 = 5a$$
$$a = 1$$

Vertical components:

The particle is in equilibrium.

The particle accelerates at 1 m s^{-2} in the direction of the 15 N force.

b $s = ut + \frac{1}{2}at^2$

$s = 0(3) + \frac{1}{2}(1)(3)^2$

$s = 4.5$

The particle travels 4.5 m in 3 s.

Example 5.3   A particle of mass 2 kg is being pulled along a rough horizontal plane by a horizontal force of magnitude 12 N. It accelerates uniformly from rest to a speed of 2 m s^{-1} in 5 seconds. Find the acceleration of the particle and hence find the coefficient of friction between the plane and the particle.

Step 1: Draw the force diagram (resolving the forces into any two perpendicular directions where necessary).

$v = u + at$

$2 = 0 + a(5)$

$a = 0.4$

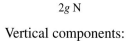

Step 2: Find the resultant force in each direction (and equate each to 0 if in equilibrium or to *ma* if accelerating).

Step 3: Solve for unknowns, using $F = \mu R$.

Horizontal components:

$F_h = ma$

$12 - F = 2(0.4)$

$F = 11.2$

$F = \mu R$

$\mu = \dfrac{F}{R} = \dfrac{11.2}{19.6} = 0.57 \text{ (2 s.f.)}$

Vertical components:

$F_v = 0$

$R - 2g = 0$

$R = 19.6$

The normal reaction is 19.6 N.

5.2   Vertical motion and motion on an inclined plane

Model, in suitable circumstances, the motion of a body moving vertically or on an inclined plane as motion with constant acceleration, and understand any limitations of this model.

Vertical motion

When the net force acts vertically on an object it will accelerate in the vertical plane according to $\mathbf{F} = ma$.

Example 5.4 A stone of mass 2 kg is attached to the lower end of a string hanging vertically. The particle is raised and moves with an acceleration of $5\ \mathrm{m\ s^{-2}}$. Find the tension in the string.

Step 1: Draw the force diagram resolving the forces into any two perpendicular directions.

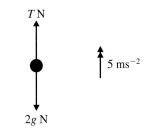

T N

$5\ \mathrm{ms^{-2}}$

$2g$ N

Note:
Let the tension in the string be *T* N.

Note:
Forces are all vertical and so do not need to be resolved into two directions.

Step 2: Find the resultant force in each direction (and equate each to 0 if in equilibrium or to *ma* if accelerating).

Step 3: Solve for unknowns.

No horizontal components.

Vertical components:

$$F_v = ma$$
$$T - 2g = 2(5)$$
$$T = 2g + 10$$
$$= 29.6$$

The tension in the string is 29.6 N.

Motion on an inclined plane

You can also apply $\mathbf{F} = ma$ when a particle is on an inclined plane if there is a resultant force that will cause the particle to accelerate up or down the plane. The resultant force parallel to the plane will be referred to as $F_{\parallel}$.

Example 5.5 A block of mass 13 kg is released from rest on a rough plane inclined at an angle of $\theta°$, where $\sin \theta° = \frac{5}{13}$. It slides down the plane and reaches a speed of $4\ \mathrm{m\ s^{-1}}$ in 2 seconds. Using the equations of motion, find the coefficient of friction between the plane and the block.

Drawing a force diagram:

The components parallel and perpendicular to the plane are:

Step 1: Draw the force diagram resolving the forces into any two perpendicular directions.

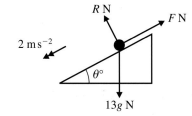

R N   *F* N
$2\ \mathrm{m\,s^{-2}}$
$\theta°$
$13g$ N

R N   *F* N
$13g \sin \theta°$ N $\theta°$
$13g \cos \theta°$ N

Note:
Use $v = u + at$ with $v = 4$, $u = 0$, $t = 2$.
$4 = 0 + 2a$
$a = 2$
The acceleration is $2\ \mathrm{m\ s^{-2}}$.

Step 2: Find the resultant force in each direction (and equate each to 0 if in equilibrium or to *ma* if accelerating).

Step 3: Solve for unknowns, using $F = \mu R$.

Parallel to the plane:

$$F_{\parallel} = ma$$
$$13g \sin \theta° - F = 13(2)$$
$$5g - F = 26$$
$$F = 5g - 26$$
$$= 23$$
$$F = \mu R$$

$$\mu = \frac{F}{R} = \frac{23}{12g} = 0.20\ldots$$

Perpendicular to the plane: (in equilibrium)

$$R - 13g \cos \theta° = 0$$
$$R - 12g = 0$$
$$R = 12g$$

Note:
$\sin \theta° = \frac{5}{13}$
$\cos \theta° = \frac{12}{13}$

Note:
The particle is in equilibrium perpendicular to the plane.

The coefficient of friction is 0.20 (2 s.f.).

1 Find the acceleration produced when a particle of mass m kg is acted on by a resultant horizontal force of F N when:

   a $F = 8, m = 2$ **b** $F = 12, m = 6$ **c** $F = 0.5, m = 0.1$ **d** $F = \frac{1}{2}, m = 0.9$.

2 Find the resultant horizontal force that acts on a particle of mass m kg to produce an acceleration of a m s^{-2} when:

   a $m = 3, a = 2$ **b** $m = 0.55, a = 3$ **c** $m = 0.3, a = 0.9$.

3 A particle of mass 3 kg is being pulled across a rough horizontal plane by a horizontal force of 10 N. The coefficient of friction between the particle and the plane is 0.1.

   a Find the magnitude of the acceleration of the particle.

   b Given also that the particle is initially at rest, find the distance moved by the particle in the first 4 seconds.

4 A truck of mass 1500 kg is brought to rest in 5 seconds from a speed of 15 m s^{-1} on a smooth horizontal plane. Neglecting air resistance, find the braking force required to achieve this.

 5 The driving force produced by the engine of a car of mass one tonne causes a car to accelerate uniformly from rest to a speed of 8 m s^{-1} in 6 seconds along a rough horizontal road. If the coefficient of friction between the car and the plane is 0.3, find the driving force of the car.

6 A wall is raised vertically upwards, by a rope. It starts from rest and travels a vertical distance of 4 m in 6 seconds. If the tension in the rope is 400 N, find the mass of the wall. What assumptions have you made when modelling this situation?

7 A boy of mass 30 kg slides down a smooth plane inclined 45° to the horizontal. Find the acceleration of the boy.

8 Repeat question **5**, with the car moving up a rough plane inclined 30° to the horizontal.

 9 A particle of mass 6.5 kg is in limiting equilibrium on a rough plane inclined at $\theta°$ to the horizontal where $\sin \theta° = \frac{5}{13}$. Find the coefficient of friction between the plane and the particle. A horizontal force of 78 N is now applied to the particle so that it starts to accelerate up the plane. Find the acceleration of the particle and hence find how far up the plane it moves in 2 seconds.

SKILLS CHECK **5A EXTRA** is on the CD

5.3 Connected particles and pulleys

Solve simple problems which may be modelled as the motion of two particles, connected by a light inextensible string which may pass over a fixed smooth peg or light pulley.

Horizontal motion of connected particles

When a particle is connected to another particle by a string then you can analyse the motion of the two **connected** particles by using Newton's third law. Both particles will experience the same magnitude of force, T N, but in opposite directions.

Note:
Because they are connected they will both have the same acceleration.

To analyse this motion apply $\mathbf{F} = ma$ to each particle, separately.

Recall:
What assumptions must be made?

Example 5.6 Two particles of masses 5 kg and 7 kg are connected by an inextensible string. The particle of mass 7 kg is being pulled by a horizontal force of 70 N along a rough, horizontal surface. Given that the coefficient of friction between each particle and the surface is 0.25, find the acceleration of the system and the tension in the string. What assumption have you made about the string?

Step 1: Draw the force diagram resolving the forces into any two perpendicular directions.

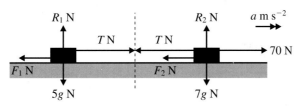

Step 2: Find the resultant force in each direction (and equate to 0 if in equilibrium, to *ma* if accelerating).

5 kg particle:

$F_v = 0$ (in equilibrium)

$R_1 - 5g = 0$ ①

$F_h = ma$

$T - F_1 = 5a$ ②

7 kg particle:

$F_v = 0$ (in equilibrium)

$R_2 - 7g = 0$ ③

$F_h = ma$

$70 - T - F_2 = 7a$ ④

Step 3: Solve for unknowns, using $F = \mu R$.

$R_1 = 49$ from ①

$F_1 = \mu R_1 = 0.25(49) = 12.25$

$R_2 = 68.6$ from ③

$F_2 = \mu R_2 = 0.25(68.6) = 17.15$

Substitute into ②

$T - 12.25 = 5a$ ⑤

Substitute into ④

$70 - T - 17.15 = 7a$ ⑥

Add equations ⑤ and ⑥: $40.6 = 12a$

$a = 3.383\ldots = 3.4$ (2 s.f.)

$T = 5a + 12.25$ from ⑤

$= 29.166\ldots = 29.2$ (2 s.f.)

Both particles accelerate at 3.4 m s^{-2} with a tension of 29 N in the connecting string.

We have assumed that the string is light, otherwise there would have to be a vertical component for the weight of the string.

Vertical motion of connected particles

This method also applies to particles that are connected vertically.

Example 5.7 A light, inextensible string connects two bricks of equal mass 5 kg, one above the other. The system is lowered by a tow bar, which is attached to the topmost brick. It takes 10 seconds for the bricks to travel a vertical distance of 15 m, starting from rest. Find the acceleration of the bricks and use this to find the tensions in the string and the tow bar.

Step 1: Draw the force diagram resolving the forces into any two perpendicular directions.

Use $s = ut + \frac{1}{2}at^2$

$15 = 0(10) + \frac{1}{2}a(10)^2$

$a = 0.3$

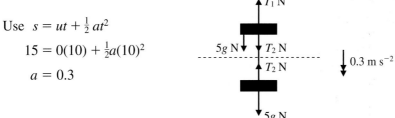

Step 2: Find the resultant force in each direction (equate to 0 if in equilibrium, or to *ma* if accelerating).

Top brick:

$F_v = ma$

$T_2 + 5g - T_1 = 5(0.3)$ ①

Bottom brick:

$F_v = ma$

$5g - T_2 = 5(0.3)$ ②

No horizontal components.   No horizontal components.

Step 3: Solve for unknowns.

$$T_2 = 5g - 1.5 \qquad \text{from ②}$$
$$= 47.5$$
$$T_1 = T_2 + 5g - 1.5 \qquad \text{from ①}$$
$$= 95$$

Note:
Alternatively equations ① and ② could be solved simultaneously.

The tension in the tow bar is 95 N and the tension in the string is 47.5 N.

Particles connected by pulleys

The motion of two particles that are connected by a light, inextensible string passing over a smooth fixed pulley can be similarly analysed. **F** = *ma* is applied separately to each particle. As the particles are connected by a light, inextensible string the magnitude of the acceleration is the same for both particles (but it acts in opposite directions). As the pulley is smooth, the tension in the string is the same throughout the string.

Example 5.8 Particles of mass *m* kg and 3*m* kg are connected by a light, inextensible string, which passes over a smooth fixed pulley. Find, in terms of *g*, the acceleration of the system and the force exerted on the pulley.

Given that the system is released from rest, find the distance moved by one of the particles in 3 s, assuming the particle does not reach the pulley.

Step 1: Draw the force diagram resolving the forces into any two perpendicular directions.

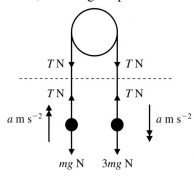

Note:
Remember to imagine a line half way down the string as shown.

Note:
Mass *m* moves up whilst mass 3*m* moves down.

Step 2: Find the resultant force in each direction (equate to 0 if in equilibrium, or to *ma* if accelerating).

For mass *m*:   For mass 3*m*:

$$T - mg = ma \qquad ① \qquad 3mg - T = 3ma \qquad ②$$

No horizontal components.   No horizontal components.

Note:
Define the direction of acceleration as positive.

Step 3: Solve for unknowns.

① + ②:
$$2mg = 4ma$$
$$a = \frac{2g}{4} = \frac{g}{2}$$

The acceleration of the system is $\frac{g}{2}$ m s^{-2}.

The force exerted on the pulley is 2*T* N, where
$$T = ma + mg \qquad \text{from ①}$$
$$= \frac{3mg}{2}$$

So, the force exerted on the pulley is 3*mg* N downwards.

For the distance travelled in 3 seconds use $s = ut + \frac{1}{2}at^2$

$$s = 0(3) + \frac{1}{2}\left(\frac{g}{2}\right)3^2 = 22.1 \ (3 \text{ s.f.})$$

Recall:
Equations of motion in Chapter 4.

So, the distance travelled is 22.1 m.

Horizontal and vertical motion of particles connected by pulleys

Sometimes a pulley can connect two particles, one of which is resting on a plane and the other hanging freely.

Example 5.9 Two particles A and B of masses 0.5 kg and 0.7 kg respectively are connected by a light, inextensible string. Particle A lies on a rough horizontal table 8 m from a smooth peg at the edge of the table. The string passes over the peg and particle B hangs freely 2 m from the ground. The coefficient of friction between particle A and the horizontal surface is 0.2. The system is released from rest. Find:

a the acceleration of the system

b the time taken for B to reach the ground

c the distance that A travels along the table, after B reaches the ground, before it comes to rest.

> **Note:**
> A smooth peg can be treated in exactly the same way as a smooth pulley.

Step 1: Draw the force diagram resolving the forces into any two perpendicular directions.

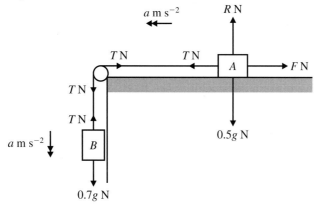

> **Note:**
> Apply $\mathbf{F} = ma$ horizontally for particle A and vertically for particle B. Define the direction of motion as positive.

Step 2: Find the resultant force in each direction (equate to 0 if in equilibrium, or to ma if accelerating).

For 0.5 kg mass:
Horizontal motion:
$$F_h = ma$$
$$T - F = 0.5a \qquad \text{①}$$

Vertical motion:
$$F_v = 0 \text{ (in equilibrium)}$$
$$R - 0.5g = 0 \qquad \text{②}$$

For 0.7 kg mass:
No horizontal motion.

Vertical motion:
$$F_v = ma$$
$$0.7g - T = 0.7a \qquad \text{③}$$

> **Note:**
> Consider motion whilst particle B hangs freely.

Step 3: Solve for unknowns using $F = \mu R$.

a $R = 4.9$ from ②
$$F = \mu R = 0.98$$
Substitute into ① $T - 0.98 = 0.5a \qquad \text{④}$
$$5.88 = 1.2a$$
$$a = 4.9$$

The acceleration of the particles as B falls is 4.9 m s^{-2}.

> **Note:**
> Substitute value for R into ① then add ③ and ④.

b $s = ut + \frac{1}{2}at^2$ (where $s = 2$, $u = 0$, $a = 4.9$, $t = ?$)
$$2 = 0(t) + \tfrac{1}{2}(4.9)t^2$$
$$t = \sqrt{\frac{4}{4.9}} = 0.903\ldots$$

Particle B falls for 0.90 s (2 s.f.) before hitting the ground.

> **Note:**
> Use the value of a to calculate the time and final velocity while the string is taut.

Step 1: Find the final velocity of the particle while the string is taut.

c Find the velocity of the particles as B hits the ground:
$$v^2 = u^2 + 2as \qquad \text{(where } s = 2, u = 0, v = ?, a = 4.9\text{)}$$
$$v^2 = 0(2) + 2(4.9)(2) = 19.6$$
$$v = 4.427\ldots$$

Find the acceleration of particle A:

You can apply $\mathbf{F} = ma$ to particle A or you can recognise that the equation will be the same as the one you did earlier but without any tension. So, you can use equation ④ but remove the tension to get the new acceleration when the particles are no longer connected:

$$-0.98 = 0.5a \qquad \text{from ④}$$
$$a = -1.96$$

Step 3: Use the equations of motion to find the further distance moved by the particle until it stops.

Find the further distance travelled by A:

$$v^2 = u^2 + 2as \qquad \text{(where } u = 4.427\ldots, v = 0, a = -1.96\text{)}$$
$$0^2 = 4.427\ldots^2 + 2(-1.96)s$$
$$s = 5$$

Particle A travels 5 m along the table after particle B hits the ground (the total distance travelled by A is $(5 + 2)\,\text{m} = 7\,\text{m}$).

Example 5.10 Two particles P and Q, of masses 10 kg and 15 kg respectively, are connected by a light, inextensible string which passes over a light, smooth pulley. Particle P rests on a smooth plane inclined at $\theta°$ to the horizontal, where $\sin \theta° = \frac{3}{5}$. Particle Q hangs vertically on the edge of the plane, 2 m above a horizontal plane. Find:

a the acceleration of the system

b the tension in the string

c the total distance that P travels up the plane, given that the string breaks after Q has travelled 1 m.

Step 1: Draw the force diagram resolving the forces into any two perpendicular directions.

Draw a force diagram:

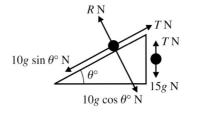

Resolve parallel to the plane for P and vertically for Q:

Step 2: Find the resultant force in each direction (and equate to 0 if in equilibrium, to ma if accelerating).

Step 3: Solve for unknowns.

Particle P:

Parallel to plane:

$$F_{||} = ma:$$
$$T - 10g \sin \theta° = 10a$$
$$T - 6g = 10a \qquad ①$$

Particle Q:

Vertical motion:

$$F_v = ma$$
$$15g - T = 15a \qquad ②$$

a $9g = 25a \Rightarrow a = \dfrac{9 \times 9.8}{25} = 3.528$

The acceleration of the system is $3.5\,\text{m s}^{-2}$ (2 s.f.).

b $T = 10a + 6g = 94.08$

The tension in the string is 94 N (2 s.f.).

Step 1: Find the final velocity of the particle while the string is taut.

c Find the velocity as the string breaks:

$$v^2 = u^2 + 2as \qquad \text{(where } s = 1, u = 0, v = ?, a = 3.5\text{)}$$

$$v^2 = 0^2 + 2 \times \frac{9g}{25} \times 1$$

$$v = \sqrt{7.056}$$

Tip:
Use accurate values in calculations to avoid rounding errors.

Step 2: Find the new acceleration of the particle when there is no tension.

Resolve the forces on *P*, parallel to the plane, after the string breaks:

$$-6g = 10a$$

$$a = -5.88$$

The acceleration of particle *P* after the string breaks is $-5.88 \, \text{m s}^{-2}$.

Recall:
Use equation ① without the tension.

Step 3: Use the equations of motion to find the further distance moved by the particle until it stops.

Find the distance moved after the string breaks:

$$v^2 = u^2 + 2as \qquad \text{(where } u = \sqrt{7.056}, v = 0, s = ?, a = -5.88\text{)}$$

$$0^2 = 7.056 - 2 \times 5.88 \times s$$

$$s = 0.6$$

So the total distance travelled by *P* up the plane is 1.6 m.

Note:
a is negative because the motion is decelerating.

SKILLS CHECK **5B: Connected particles and pulleys**

1 Two particles of mass 5 kg and 8 kg are on a rough horizontal surface and are connected by a light, inextensible string. The 8 kg mass is being pulled by a horizontal force of 39 N. The resistance to each particle is *k* times their mass, so the resistance experienced by the 5 kg mass is 5*k* N. If the acceleration of the system is $1 \, \text{m s}^{-2}$, find the tension in the connecting string and the magnitude of the resistance experienced by each particle.

 2 A dog of mass 6 kg is attached to a sleigh of mass 3 kg by a rope and pulls it along a smooth horizontal surface. The dog exerts a forward force of *F* N. The dog and the sleigh start from rest and travel a distance of 15 m in 10 seconds. Find the force with which the dog pulls the sleigh and find the tension in the rope. The dog and the sleigh now reach a smooth hill inclined at 1° to the horizontal and continue along the line of greatest slope of the plane. If the dog maintains the same pulling force find the acceleration of the dog and the sleigh up the plane. What assumptions have you made when modelling this situation?

3 A light, inextensible string connects two particles *A* and *B*. The particles hang vertically with particle *B* below particle *A*. Particle *A* has a mass of 3 kg and particle *B* has a mass of 2 kg. Particle *A* is pulled upwards by means of another string with tension 85 N. Find the acceleration of the system and the tension in the lower string.

 4 A crane lowers two cars that are connected by a chain. The topmost car has mass 1000 kg and is suspended by a rope, which is connected to the crane. The lower car has mass 800 kg and is 100 m vertically above the ground. Given that it takes 10 seconds for the lower car to reach the ground, starting from rest, find the tensions in the chain and in the rope. What assumptions have you made about the chain and the rope?

5 Particles of mass 3 kg and 5 kg are attached by a light, inextensible string, which passes over a smooth fixed pulley. The 5 kg particle is 3 m above ground. The system is released from rest. By finding the acceleration of the system, find how long it takes the 5 kg mass to reach the ground. Also, find the force exerted on the pulley.

6 Particles *P* and *Q* are attached by a string that passes over a smooth fixed pulley. Particle *P* has twice the mass of particle *Q*. They both hang 2 m above horizontal ground. The system is released from rest. Find the magnitude of the acceleration of the system. Hence, find the velocity with which *P* hits the ground. Assuming that particle Q does not reach the pulley, find the greatest height that Q reaches above the ground. How have you used the fact that the pulley is smooth? How have you used the fact that the string is light **and** inextensible?

7 Two particles A and B are connected by a light, inextensible string. Particle A has mass 6 kg and lies on a rough horizontal table. The string passes over a smooth fixed pulley at the edge of the table and particle B of mass 5 kg hangs vertically at the other end of the string, 2 m above horizontal ground. The system is released from rest and it takes 5 seconds for particle B to hit the ground. Find the coefficient of friction between the plane and particle A.

8 A particle P of mass 2.5 kg which is at rest on a smooth inclined plane of angle 30° is connected to particle Q of mass 3 kg by a light, inextensible string which lies along a line of greatest slope of the plane and passes over a fixed smooth pulley at the top of the plane. Particle Q hangs freely 1.5 m above horizontal ground. The system is released from rest with the string taut. Find:

    a the acceleration of the system

    b the tension in the string

    c the final velocity of Q when it hits the ground

    d the total distance that P moves up the plane, given that it does not reach the pulley.

9 The diagram shows two particles P and Q of masses 0.4 kg and 2 kg respectively. Particle P rests on a rough plane inclined at angle 30° to the horizontal and is attached to particle Q by means of a light, inextensible string which passes over a smooth fixed pulley at the top of the plane as shown. Particle Q lies on a smooth plane inclined at 60° to the horizontal. The coefficient of friction between particle P and the plane is $\dfrac{1}{\sqrt{3}}$. The system is released from rest.

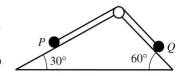

Find the acceleration of the system, given that Q travels down the plane.

SKILLS CHECK **5B EXTRA** is on the CD

Examination practice   Newton's laws of motion

1 A car is towing a trailer along a horizontal straight road using a horizontal tow-bar. The masses of the car and the trailer are 1050 kg and 200 kg respectively. The resistance to motion of the car is 850 N and the resistance to motion of the trailer is 150 N.

    i At an instant when the driving force exerted by the car is 1100 N, find

        a the acceleration of the car,

        b the pulling force exerted on the trailer.

    ii At another instant the pulling force exerted on the trailer is zero.

        a Show that the acceleration of the car is -0.75 m s^{-2}.

        b Find the driving force exerted by the car.   [OCR Jan 2004]

2

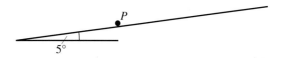

A particle P is released from rest at the top of a smooth plane, of length 5 m, which is inclined at 5° to the horizontal (see diagram). Air resistance may be neglected. Find

i   the acceleration of P down the plane,

ii   the time taken for P to reach the bottom of the plane,

iii   the speed with which P reaches the bottom of the plane.   [OCR June 2001]

3

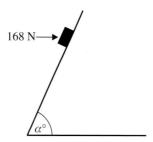

A smooth plane is inclined at $\alpha°$ to the horizontal. A block of mass 5 kg is held at rest on the plane by a horizontal force of magnitude 168 N (see diagram). Find, in either order,

i   the value of α,

ii   the magnitude of the normal force exerted by the plane on the block.

The horizontal force is now removed.

iii   Find the distance travelled by the block in 0.8 s.   [OCR June 2004]

 4

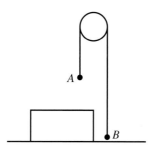

Particles A and B, of masses 0.9 kg and 0.5 kg respectively, are attached to the ends of a light inextensible string. The string passes over a smooth fixed pulley. The string is taut and the two parts which are not in contact with the pulley are vertical. The system is released from rest with A vertically above a horizontal step and with B on the floor (see diagram).

i   For the motion before A hits the step, find the acceleration of A and the tension in the string.

Particle A reaches the step 0.8 s after the system is released from rest. A then remains at rest on the step, the string becomes slack and B continues to rise.

ii   Find the greatest height above the floor reached by B.

iii   Find the time for which A is at rest on the step before starting to move upwards.   [OCR June 2004]

5

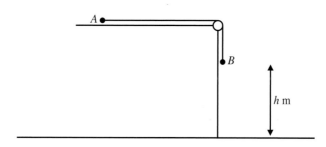

Particles A and B, each of mass 0.6 kg, are joined by a light inextensible string. The string passes over a smooth pulley at the edge of a smooth horizontal platform. A is held at rest on the platform. B hangs vertically below the pulley at a height h m above the floor, as shown in the diagram. A is released, with the string taut, and the particles start to move. There is no air resistance.

i Find the tension in the string and the acceleration of A.

ii Hence find the speed of A after it has travelled a distance of 2 m.

iii When A has moved a distance of 2 m it becomes detached from the string. From this instant B takes a further 0.2 s to reach the floor. Find the value of h.

iv Find also the total time for which B is in motion before it reaches the floor.　　　[OCR Jan 2001]

6 A sledge of mass 25 kg is on a plane inclined at 30° to the horizontal. The coefficient of friction between the sledge and the plane is 0.2.

i

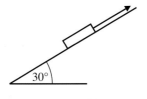

Fig. 1

The sledge is pulled up the plane, with constant acceleration, by means of a light cable which is parallel to a line of greatest slope (see Fig. 1). The sledge starts from rest and acquires a speed of 0.8 m s^{-1} after being pulled for 10 s. Ignoring air resistance, find the tension in the cable.

ii

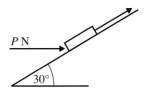

Fig. 2

On a subsequent occasion the cable is not in use and two people of total mass 150 kg are seated in the sledge. The sledge is held at rest by a horizontal force of magnitude P newtons, as shown in Fig. 2. Find the least value of P which will prevent the sledge from sliding down the plane.　　　[OCR June 2002]

7

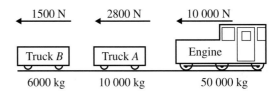

A train consists of an engine of mass 50 000 kg coupled to two trucks A and B of masses 10 000 kg and 6000 kg respectively (see diagram). The couplings are light, rigid and horizontal. The train moves along a horizontal track with a constant deceleration. The resistance to motion of the engine, truck A and truck B are 10 000 N, 2800 N and 1500 N respectively. The tension in the coupling between truck A and truck B is zero.

i By applying Newton's second law to truck B, show that the deceleration of the train is 0.25 m s^{-2}.

ii Find the tension in the coupling between the engine and truck A.

iii Determine whether the engine exerts a driving force or a braking force, and find its magnitude.　　　[OCR June 2003]

8 A particle of mass M kg slides down a rough plane which is inclined at 30° to the horizontal. The particle passes through a point A with speed 4 m s^{-1}, and 2 s later it passes through a point B with speed 9 m s^{-1}.

   i Find the acceleration of the particle.

   ii Find the distance AB.

   iii Show that the frictional force on the particle is 2.4M newtons.

   iv Find the coefficient of friction between the particle and the plane.   [OCR June 2003]

9

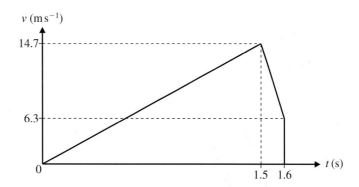

A stone falls vertically from rest, under gravity, into a drum of oil. The diagram shows the (t, v) graph for the motion of the stone, where v m s^{-1} is the stone's downwards velocity at time t seconds after it starts to fall.

   i Use the information in the diagram to describe briefly the motion of the stone after it reaches the surface of the oil.

   ii Find the height, above the surface of the oil, from which the stone falls.

   iii Find the depth of oil in the drum.

   iv Find the deceleration of the stone when $t = 1.55$.

   v The mass of the stone is 0.08 kg. Find the upward force acting on the stone, due to the oil, when $t = 1.55$.   [OCR June 2003]

  10 A board is fixed so that it makes an angle of 11° with the horizontal. A block of mass 0.2 kg is placed on the board and then set in motion with an initial speed of 2 m s^{-1} down a line of greatest slope of the board. The block comes to rest in 4 s. The coefficient of friction between the block and the board is μ. Find

   i the deceleration of the block,

   ii the frictional force on the block while the block is in motion,

   iii the value of μ.

With the block at rest on the board, the inclination of the board is gradually increased. The angle that the board makes with the horizontal is α. Find α when

   iv the block starts to slide,

   v the block is moving with acceleration $g(1 - \mu)\cos\alpha$.   [OCR Jan 2004]

6 Linear momentum

6.1 Linear momentum

Recall and use the definition of linear momentum and show understanding of its vector nature.

Momentum

The momentum of a particle of mass m kg, travelling with velocity $\mathbf{v}$ m s^{-1} has magnitude mv. Momentum is a vector quantity and is measured in units of N s (newton seconds).

Momentum $= m\mathbf{v}$ N s

If the particle has speed v m s^{-1} then the momentum is given by:

Momentum $= mv$ N s

Recall:
$1\,\text{N} = 1\,\text{kg}\,\text{m}\,\text{s}^{-2}$.

Example 6.1 A particle of mass 3 kg is travelling with velocity of 4 m s^{-1}. Calculate:

    a the momentum of the particle

    b the change in momentum if the velocity of the particle increases to 8 m s^{-1}.

Step 1: Calculate unknowns using the definition of momentum.

  a $mv = 3 \times 4 = 12$

    The initial momentum of the particle is 12 N s.

  b $mv = 3 \times 8 = 24$

    The final momentum is 24 N s.

    The change in momentum $= 24 - 12 = 12$ N s.

Note:
You must consider the direction of momentum.

Example 6.2 A particle of mass 3 kg has initial velocity of 2.5 m s^{-1}. Find the change in momentum of the particle if:

    a the final velocity of the particle is 6 m s^{-1}

    b the final velocity of the particle is -6 m s^{-1}.

Note:
A negative sign indicates opposite direction.

Step 1: Draw a diagram of the initial and final situations

a Initial motion: Final motion:

Step 2: Calculate unknowns using the definition of momentum.

Initial momentum $= mv$ Final momentum $= mv$
$= 3 \times 2.5$ $= 3 \times 6$
$= 7.5$ $= 18$

Change in momentum $= 18 - 7.5 = 10.5$ N s.

Step 1: Draw a diagram of the initial and final situations

b Initial motion: Final motion:

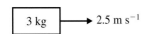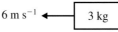

Step 2: Calculate unknowns using the definition of momentum.

Initial momentum $= mv$ Final momentum $= mv$
$= 3 \times 2.5$ $= 3 \times -6$
$= 7.5$ $= -18$

Change in momentum $= (7.5 - (-18))$ N s $= 25.5$ N s.

Note:
The direction is important. The velocity to the right is taken as positive.

Understand and use conservation of linear momentum in simple applications involving the direct collision of two bodies moving in the same straight line, before and after impact, including the case where the bodies coalesce.

Momentum is conserved between two colliding particles when there are no external forces acting. That is, **the total momentum before the collision equals the total momentum after the collision.** This is called the **principle of conservation of momentum**. You can use this relation to solve problems involving colliding particles.

Example 6.3 A particle of mass 4 kg travels horizontally with a velocity of 8 m s^{-1}. It collides with another particle of mass 3 kg, which is at rest. After the impact the 4 kg mass has a velocity of -5 m s^{-1}. Find the velocity of the 3 kg mass after the impact.

> **Note:**
> Let the speed of the 3 kg mass after the collision be v m s^{-1}.

Step 1: Draw a diagram for the initial and final situations.

Before collision:

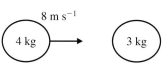

After collision:

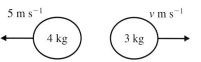

> **Note:**
> The direction of motion is very important, so choose which direction is positive at the start and stick with it.

Step 2: Calculate the total momentum before and after the collision.

Momentum before collision
$$= 4 \times 8 + 0$$
$$= 32$$

Momentum after collision
$$= 4 \times (-5) + 3v$$
$$= -20 + 3v$$

> **Note:**
> Let motion from left to right be positive.

Step 3: Calculate unknowns using the principle of conservation of momentum.

By the principle of conservation of momentum
$$32 = -20 + 3v$$
$$v = \tfrac{52}{3} = 17.33\ldots$$

> **Recall:**
> Momentum before
> = momentum after.

The final velocity of the particle is 17 m s^{-1} (2 s.f.).

Example 6.4 Two particles A and B of masses 2 kg and 1 kg respectively are moving towards each other in the same straight line with speeds $2u$ m s^{-1} and u m s^{-1}, respectively. After the impact, the particles coalesce and continue to travel in the direction of particle A before impact, with speed 5 m s^{-1}. Find the initial speeds of A and B.

> **Note:**
> When two particles coalesce, they join together and you can treat them as a single particle.

Step 1: Draw a diagram for the initial and final situations.

Before collision:

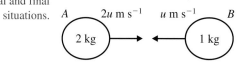

After collision:

> **Note:**
> Motion from left to right is positive.

Step 2: Calculate the total momentum before and after the collision.

Momentum before collision
$$= 2(2u) - u$$
$$= 3u$$

Momentum after collision
$$= 15$$

> **Tip:**
> It is useful to simplify each equation at this stage of the calculation.

Step 3: Calculate unknowns using the principle of conservation of momentum.

By the principle of conservation of momentum
$$3u = 15$$
$$u = 5$$

The initial speeds of A and B are 10 m s^{-1} and 5 m s^{-1}, respectively.

Gun firing a bullet

When a gun fires a bullet the initial momentum is 0 N s, assuming there is no initial movement. But on firing the bullet there is a **recoil** in the gun which is in the opposite direction to the motion of the bullet, according to the conservation of momentum.

Recall:
Initial momentum = final momentum.

Example 6.5 A bullet is fired by a gun, which is 3 kg heavier than the bullet. Given that, after the shot is fired, the bullet travels in a straight line with velocity 200 m s^{-1} and the gun recoils in the opposite direction with velocity 5 m s^{-1}, find the mass of the bullet and the mass of the gun.

Step 1: Draw a diagram for the initial and final situations.

5 m s^{-1} 200 m s^{-1}

Note:
Let the mass of the bullet be m kg. Then the mass of the gun is $(m + 3)$ kg.

Step 2: Calculate the total momentum before and after the collision.

Momentum before firing
$= 0 \text{ N s}$

Momentum after firing
$= (3 + m)(-5) + 200m$
$= (195m - 15) \text{ N s}$

Note:
Motion from left to right is positive.

Step 3: Calculate unknowns using the principle of conservation of momentum.

By the principle of conservation of momentum
$$0 = 195m - 15$$
$$m = 0.0769\ldots$$

So the mass of the bullet is 77 g (2 s.f.) and the mass of the gun is $(3 + 0.08) \text{ kg} = 3.08 \text{ kg}$ (3 s.f.).

Note:
Because the value of m is so small it is better to convert it to grams.
$1 \text{ kg} = 1000 \text{ g}$

Jerk in a string

If two particles are connected by a (light) string, which is initially slack, and one particle is given an initial velocity, then when the string becomes taut there will be a jerk in the string. After the jerk, **both particles will move off with the same velocity.**

Note:
Let the particle's velocity be $v \text{ m s}^{-1}$.

Example 6.6 Two identical particles P and Q, each of mass m kg, lie on a smooth, horizontal table. They are connected at either end of a light, inextensible string which initially is slack. Particle Q is projected away from P with an initial velocity of 8 m s^{-1}. Find the common velocity of the particles after the string jerks

Step 1: Draw a diagram for the initial and final situations.

Before collision: $+$ After collision:

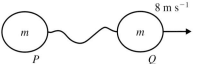 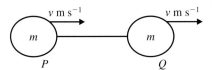

Note:
Only one particle has a velocity before the jerk but they both have the same velocity after the jerk.

Step 2: Calculate the total momentum before and after the collision.

Momentum before jerk
$= 8m$

Momentum after jerk
$= mv + mv$
$= 2mv$

Note:
Let motion from left to right be positive.

Step 3: Calculate unknowns using the principle of conservation of momentum.

By the principle of conservation of momentum:
$$8m = 2mv$$
$$v = 4$$

So the combined speed after the string becomes taut is 4 m s^{-1}.

1 Calculate the momentum for a particle of mass m kg and travelling with velocity v m s^{-1} when:

   a $m = 2, v = 4$ **b** $m = 0.1, v = 18$ **c** $m = \frac{2}{3}, v = 15$.

2 Find the magnitude of the change in momentum of a particle of mass 8 kg that changes its speed from 2 m s^{-1} to:

   a 14 m s^{-1} in the same direction   **b** 14 m s^{-1} in the opposite direction.

In questions **3**, **4** and **5**, two particles collide as shown. The diagrams show the situation before and after the impact. Find the unknown initial velocity u or the unknown final velocity v as appropriate:

3 Before collision: After collision:

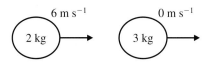 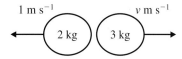

4 Before collision: After collision:

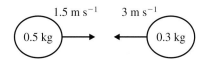 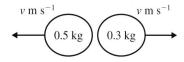

5 Before collision: After collision:

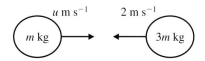

6 A car of mass 800 kg travelling on a smooth road with speed 6 m s^{-1} collides and coalesces with a stationary car of mass 500 kg. Find the speed of the combined cars after the collision. Also find the magnitude of the change in momentum of the moving car.

7 A gun of mass 3 kg fires a bullet of mass 20 g. If the gun recoils at 2 m s^{-1}, find the speed of the bullet after the shot is fired. The bullet drives into a wall with this speed and stops after it has driven 2 cm into the wall in a horizontal direction. Find the magnitude of the resistance provided by the wall.

8 Two particles A and B of mass M kg and $2M$ kg respectively are connected by a light, inextensible string, which is initially slack. Particle B is projected away from particle A with speed u m s^{-1}. Find u if the combined speed of the two particles after the string becomes taut is 4 m s^{-1}.

SKILLS CHECK **6A EXTRA is on the CD**

Examination practice   Linear momentum

1 Two particles P and Q have masses 0.05 kg and 0.03 kg respectively. The particles are moving towards each other, P with speed 1.2 m s^{-1} and Q with speed 1.3 m s^{-1}, when they collide directly. Particle P is brought to rest by the collision. Find the speed of Q immediately after the collision.

[OCR Jan 2002]

2

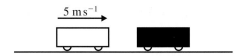

Two wagons, each of unloaded mass 1000 kg, are free to travel on a straight horizontal track. One of the wagons carries a load of mass m kg and the other is empty. The empty wagon is travelling at 5 m s^{-1} when it runs into the loaded wagon which is stationary (see diagram). Immediately after the collision the empty wagon continues in the same direction with speed 0.5 m s^{-1}, and the loaded wagon starts to move with speed 1.5 m s^{-1}. Find the value of m.   [OCR June 2001]

3

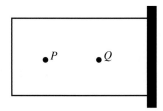

The diagram shows two particles P and Q, of masses 0.3 kg and 0.2 kg respectively, which are free to move on a smooth horizontal table. A vertical wall is situated at one end of the table. The particles are moving directly towards the wall in the same straight line, P with speed 3.2 m s^{-1} and Q with speed 1.2 m s^{-1}, when they collide.

i   Given that P continues to move towards the wall after the collision, with speed 2 m s^{-1}, find Q's speed after the collision.

Q hits the wall and rebounds from it with speed 2 m s^{-1}. The particles subsequently collide again, after which they move in opposite directions, P with speed v m s^{-1} and Q with speed $4v$ m s^{-1}.

ii Find the value of v.   [OCR Nov 2003]

4

Three uniform spheres A, B and C have masses 0.3 kg, 0.4 kg and m kg respectively. The spheres lie in a smooth horizontal groove with B between A and C. Sphere B is at rest and spheres A and C are each moving with speed 3.2 m s^{-1} towards B (see diagram). Air resistance may be ignored.

i A collides with B. After this collision A continues to move in the same direction as before, but with speed 0.8 m s^{-1}. Find the speed with which B starts to move.

ii B and C then collide, after which they both move towards A, with speeds of 3.1 m s^{-1} and 0.4 m s^{-1} respectively. Find the value of m.   [OCR June 2002]

5 Particles A and B, of masses 0.15 kg and 0.2 kg respectively, are free to move on a horizontal surface. Air resistance may be ignored. At a particular instant A is moving with speed 2 m s^{-1} towards B, which is stationary at a point 4 m from A. Particle A collides directly with particle B.

i   It is given that the horizontal surface is smooth and that A is brought to rest by the collision. Find the speed of B immediately after the collision.

ii It is given instead that the coefficient of friction between A and the surface is 0.05. A is again brought to rest by the collision. Find the speed of B immediately after the collision.   [OCR Jan 2001]

6 Two small blocks A and B have masses 0.36 kg and 0.32 kg respectively. Block B is stationary on an ice rink 17 m from a boundary wall. Block A is moving at 8 m s^{-1} at right angles to the boundary wall when it strikes B. Block A continues in the same direction, and its speed immediately after the collision is 2.4 m s^{-1}.

i Find the speed with which B starts to move.

ii Given that B moves with constant retardation, and reaches the boundary wall with speed 5.6 m s^{-1}, find the coefficient of friction between B and the surface of the ice.

Block B rebounds from the wall in the reverse direction and comes to rest 4 m from the wall. [You may assume that A and B do not collide again.]

iii Find the change in B's momentum as a result of its impact with the wall.   [OCR June 2004]

7 A smooth plane is inclined at 3° to the horizontal.

i Show that a particle, moving freely along a line of greatest slope of the plane, has a downward acceleration of approximately 0.513 m s^{-2}.

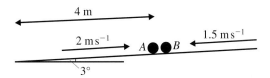

Two particles A and B, of masses 0.3 kg and 0.2 kg respectively, are moving freely along a line of greatest slope of the plane when they collide at a point 4 m from the bottom of the plane. Immediately before the collision A is moving up the plane with a speed of 2 m s^{-1} and B is moving down the plane with a speed of 1.5 m s^{-1} (see diagram). Immediately after the collision A is instantaneously at rest. Find

ii the time taken by A, from the instant of the collision, to reach the bottom of the plane,

iii the speed with which B starts to move up the plane after the collision,

iv the total distance travelled by B, from the instant of the collision until the instant it reaches the bottom of the plane.   [OCR Jan 2003]

8

Three particles A, B and C have masses 0.2 kg, m kg and 0.3 kg respectively. The particles are free to move on a smooth horizontal table. Initially the particles lie in a straight line, with B and C at rest and A moving directly towards B with speed u m s^{-1} (see diagram). After A's collision with B, the speed of A is $\frac{1}{2}u$ m s^{-1} in the same direction as before. When B collides with C, the particles B and C coalesce and begin to move with speed $\frac{1}{4}u$ m s^{-1}. Find the value of m, and find also, in terms of u, the speed of B before it collides with C.   [OCR June 1998]

9

Two particles, of masses x kg and 0.1 kg, are moving towards each other in the same straight line and collide directly. Immediately before the impact, the speeds of the particles are 2 m s^{-1} and 3 m s^{-1} respectively (see diagram).

i Given that both particles are brought to rest by the impact, find x.

ii Given instead that the particles move with equal speeds of 1 m s^{-1} after the impact, find the three possible values of x.   [OCR Specimen]

Practice exam paper

Answer **all** questions.

Time allowed: 1 hour 30 minutes

A calculator **may** be used in this paper.

1 A bolt of mass 0.5 kg is dropped from a building, and reaches the ground after falling 40 metres. Calculate its momentum when it hits the ground. *(3 marks)*

2 A car moving in a straight line at 10 m s^{-1} accelerates for 6 s at 2.5 m s^{-2}. Find the distance the car travels while accelerating and calculate its final speed. *(5 marks)*

3

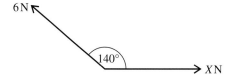

Forces of magnitude 6 N and X N act on an object. The angle between the direction of the two forces is 140°. The resultant of the two forces acts at an angle of 30° to the force of magnitude X N. Show that X is approximately 11.3 and find the magnitude of the resultant force. *(7 marks)*

4

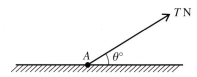

The diagram shows an object A of weight 10 N at rest on a rough horizontal floor. The coefficient of friction between A and the floor is 0.7. A force of T N is exerted on A by a rope which makes an angle $\theta°$ with the floor. Find the least value of T needed to move A when

a $\theta = 90$, *(1 mark)*

b $\theta = 0$, *(3 marks)*

c $\theta = 30$. *(5 marks)*

5

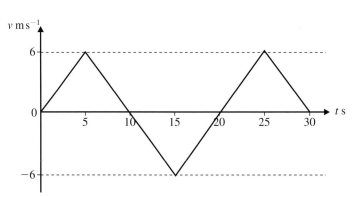

A sportsman's training regime includes running in a straight line between two points A and B. When he reaches one end of the line, he turns and runs to the other end. The (t, v) graph shows information for the first 30 s of this exercise. When $t = 0$, the sportsman is at A.

a State the time he requires to run the full length of the line. *(1 mark)*

b State the sportsman's minimum velocity, and give his position and direction of travel at this instant. *(3 marks)*

c Calculate the distance *AB*. *(3 marks)*

d Calculate his initial acceleration. *(1 mark)*

e Give two ways in which the graph would alter when the sportsman became tired. *(2 marks)*

6

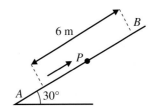

A particle *P* of mass 0.2 kg is projected from a point *A* directly up a rough slope inclined at an angle of 30° to the horizontal. The initial speed of *P* is 9 m s^{-1}. *P* comes to rest at the point *B*, 6 m from *A*.

a Show that the deceleration of *P* is 6.75 m s^{-2}. *(2 marks)*

b Calculate the normal force between *P* and the slope, and find the coefficient of friction between *P* and the slope. *(6 marks)*

c Determine whether the particle remains at rest at *B*, or moves back down the slope towards *A*. *(2 marks)*

7

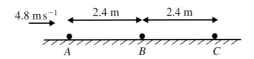

Particles *A*, *B* and *C* lie in a straight line on a smooth horizontal table. *A* is projected towards *B* with speed 4.8 m s^{-1}, with *B* and *C* at rest (see diagram). After *A* and *B* collide, they move apart with the same speed as each other.

a Given that the masses of *A* and *B* are 0.2 kg and 0.8 kg respectively, find the speed and direction of motion of *A* after the collision. *(4 marks)*

b *B* strikes *C* and is brought to rest. Given that *C* has speed 0.4 m s^{-1} after the collision, find the mass of *C*. *(3 marks)*

c Verify that the final total momentum of the three particles is equal to the initial momentum of *A*. *(3 marks)*

d Initially *AB* = *BC* = 2.4 metres. Calculate the distance between *A* and *C*, 4 s after *A* is set in motion. *(4 marks)*

8 After an aircraft *A* lands, its velocity is $64 - \dfrac{t^2}{25}$ m s^{-1}, where *t* s is the time after touchdown.

a State the velocity of *A* when it lands, and find the time taken for it to come to rest. *(3 marks)*

b Calculate the acceleration of *A* when *t* = 4, and comment on its sign. *(4 marks)*

c *A* is unable to abandon its landing once its speed has dropped below 55 m s^{-1}. Calculate the distance travelled by *A* while its speed reduces from 55 m s^{-1} to zero. *(7 marks)*

Answers

SKILLS CHECK 2A (page 7)

1. **a** 5 km, 307°   **b** 5.73 km, 282°   **c** 4.53 km, 314°
2. **a** $\mathbf{a} + \mathbf{b}$ **b** $\mathbf{a}$ **c** $\frac{1}{2}\mathbf{a}$
   d $\frac{1}{2}\mathbf{a} + \mathbf{b}$ **e** $-\mathbf{a} - \frac{1}{2}\mathbf{b}$ **f** $\frac{1}{2}(\mathbf{b} - \mathbf{a})$
3. **a** $\mathbf{b} - \mathbf{a}$ **b** $\mathbf{a} - \mathbf{b}$
   c $\frac{3}{4}(\mathbf{b} - \mathbf{a})$ **d** $\frac{1}{4}(\mathbf{a} + 3\mathbf{b})$
4. **a** 6.7 N, 206.6°   **b** 7.1 N, 135.0°   **c** 20.6 N, 29.05°
   d 3.6 N, 33.7°   **e** 5.8 N, 301.0° (or −59.0°)
5. **a** 9.8 N, 1.7 N   **b** 24.1 N, 6.5 N   **c** 3.1 N, 34.9 N
6. **a** 2.5 N, 4.3 N   **b** 17.3 N, 10 N   **c** 111.1 N, 93.2 N
7. **a** 7.8 N   **b** 8.7 N
8. **a** 16.2 N, 4.0°   **b** 98.5 N, 8.54°

Exam Practice 2 (page 9)

1. 3.8 N, 9.7°
2. 13 N, 11°
3. **i a** 2.4 N   **b** 1.7 N   **ii** 3.0 N   **iii** 36°
4. **a** 14 N   **b** 49°
5. **i** 23.4 N
   ii Weight of particle = 24.5 N vertically downwards > 23.4 N. So the resultant force, and hence the acceleration, will be vertically downwards

SKILLS CHECK 3A (page 18)

1. **a** 10 N, 8.45 N   **b** 18.8 N, 12.3 N
2. 11.7 N, 190°
3. **a** 1.96 N, 7 N   **b** 29.4 N, 14 N   **c** 88.2 N, 24 N
4. **a** 10.4 N, 43 N   **b** 13.9 N, 0.91 kg   **c** 6.93 N, 5.36 N
5. **a** 150 N   **b** 0.01
6. **a** 24.8 N, 0.78   **b** 1.36 N, 5.25 N   **c** 0.97, 1.95 kg
7. 73.6 N, 69.1 N
8. 12.3 N, 1.61 kg
9. $\frac{12}{13}mg$, $\frac{5}{13}mg$, $\frac{5}{12}$
10. **a** 27.4 N   **b** 1.97 N
11. **a** 44.5 N   **b** 5.57 N
12. 5180 N, 13 300 N

Exam Practice 3 (page 19)

1. **i** $\theta = 37$ **ii** $P = 6$ **iii** 8 N to the right
2. **i** 4.9 N   **ii** 1.96 N   **iii** 0.4
3. **i** $x = 30.5$, $P = 13$
   ii 13 N act an angle of 30.5° above 11.2 N force
4. **i** 1040 N   **ii** 0.62   **iii** 2080 N
5. **i** 840 N   **ii** 0.36
6. 40 N
7. **iii** 0.735 N, 0.127 N

SKILLS CHECK 4A (page 30)

1. **a**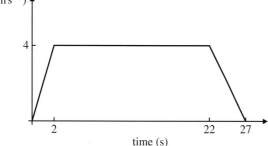

   b 2 m s⁻², −$\frac{4}{5}$ m s⁻²   **c** 94 m

2. **a**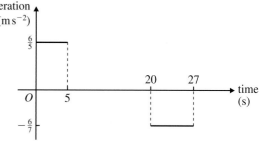

   b 7 s   **c** 126 m
   d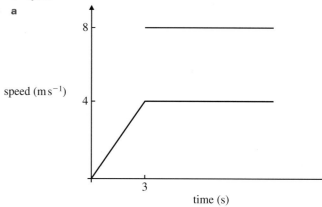

   e $4\frac{2}{3}$ m s⁻¹
3. **a**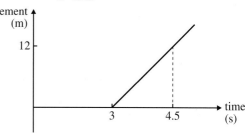

   b 4.5 s   **c** 12 m
   d

   displacement (m)

4. **a** −3 m   **b** 10 m s⁻¹   **c** 24 m s⁻²
5. **a** $(3t^2 - 7t + 2)$ m s⁻¹   **b** $t_1 = \frac{1}{3}$, $t_2 = 2$ **c** −5 m s⁻², 5 m s⁻²
6. **a** $(3t^2 - 10t + 4)$ m s⁻¹   **b** 4 m s⁻¹   **c** $(t^3 - 5t^2 + 4t)$ m
   d −4 m   **e** 6 m   **f** 1, 4 seconds
7. **a** $(-\frac{1}{10}t^2 + U)$ m s⁻¹, 40   **b** $(-\frac{1}{30}t^3 + 40t)$ m   **c** 80 m s⁻¹
8. **a** $\frac{14}{3}$ m   **b** $(4 - 4t)$ m s⁻²
   c 2 m s⁻¹   **d** $\frac{10}{3}$ m
9. **a** 5   **b** 1 s, 4 s
   c $(\frac{1}{3}t^3 - \frac{5}{2}t^2 + 4t)$ m   **d** $\frac{13}{6}$ m
10. **a** 191 m   **b** 19.4 s   **c** −1728
    d 4 s   **e** −192 m s⁻¹

SKILLS CHECK 4B (page 35)

1. **a** 2 m s⁻²   **b** 4 m
2. $\frac{25}{18}$ m s⁻²
3. 9 m s⁻¹

4 **a** $\frac{2}{3}$ s **b** No air resistance, car is a particle
5 50 m
6 **a** 13.9 s **b** 965 m
7 **a** 2.5 m s^{-2} **b** 140 m
8 **a** 4 m **b** 4.674 s **c** 10.8 m

SKILLS CHECK 4C (page 38)

1 8.4 m s^{-1}, 36.4 m
2 40 m, 5.7 s, 4.04 s
3 4.04 s; book is a particle, no air resistance, gravity constant, book starts from rest
4 **a** 14 m **b** 3.12 s
5 31.3 m s^{-1}
6 **a** 5.63 s **b** 155.3 m
7 **b** 6.38 m

Exam Practice 4 (page 39)

1 **i** $v = 10t - \frac{3}{2}t^2$, $a = 10 - 3t$ **ii** 14 m s^{-1}
2 **i** R **ii** Q **iii** 3 m s^{-1} **iv** 30 m **v** 2 m s^{-1}
3 **i** 0.12 m s^{-2} **ii** 5 m s^{-1}
4 **i** $v = -\frac{1}{20}t^2 + V$ **ii** 5 **iii** $\frac{100}{3}$ m
5 **i** 0.14 upwards, 1 downwards **ii** 0.86 s **iii** 1.2 m
6 **i** 0.15 m s^{-2} **ii** 150 m **iii** 20 s
7 **i** 2 m s^{-1} **ii** 4 s **iii** 1.7 m s^{-1} (1 d.p.)
8 **i** 29.4 **ii** 44.1 m **iii** 4.2 s
9 **i** $0.6t + 4$ **ii** 24.1 m
10 **i** 40 m **ii** 14 m s^{-1} **iii** 1.4 s **iv** 2.9 s

SKILLS CHECK 5A (page 45)

1 **a** 4 m s^{-2} **b** 2 m s^{-2} **c** 5 m s^{-2} **d** 0.56 m s^{-2}
2 **a** 6 N **b** 1.65 N **c** 0.27 N
3 2.35 m s^{-2}, 18.8 m
4 4500 N
5 4273 N
6 39.9 kg. Wall treated as a particle, neglect air resistance, rope is light, inextensible.
7 6.9 m s^{-2}
8 8780 N
9 $\frac{5}{12}$, 1.62 m s^{-2}, 3.23 m

SKILLS CHECK 5B (page 50)

1 15 N, 10 N, 16 N
2 2.7 N, 0.9 N, 0.13 m s^{-2}; the rope is light and inextensible, and there is no air resistance.
3 7.2 m s^{-2}, 34 N
4 6240 N, 14 040 N; the chain is light and the rope is light and inextensible.
5 2.45 m s^{-2}, 1.56 s, 73.5 N
6 3.27 m s^{-2}, 3.61 m s^{-1}, 4.67 m; tensions are the same in the string either side of the pulley; there is no component weight for the string; there is no force used in extending the string.
7 0.803
8 **a** 3.12 m s^{-2} **b** 20.0 N **c** 3.06 m s^{-1} **d** 2.45 m
9 5.44 m s^{-2}

Exam Practice 5 (page 51)

1 **i** **a** 0.08 m s^{-2} **b** 166 N **ii** **b** 63 N
2 **i** 0.85 m s^{-2} **ii** 3.4 s **iii** 2.9 m s^{-1}
3 **i** 74 **ii** 175 N **iii** 3 m
4 **i** 2.8 m s^{-2}, 6.3 N **ii** 1.2 m **iii** 0.46 s
5 **i** 2.9 N, 4.9 m s^{-2} **ii** 4.4 m s^{-1} **iii** 2.7 **iv** 1.1 s
6 **i** 170 N **ii** 580 N
7 **ii** 300 N **iii** braking force, 2200 N
8 **i** 2.5 m s^{-2} **ii** 13 m **iv** 0.28
9 **i** Uniformly decelerates from speed of 14.7 m s^{-1} to speed of 6.3 m s^{-1} in 0.1 s (from $t = 1.5$ to $t = 1.6$), and then comes to an immediate halt at $t = 1.6$, where it has reached the base of the oil drum
  ii 11.0 m **iii** 1.1 m **iv** 84 m s^{-2} **v** 7.5 N
10 **i** 0.5 m s^{-2} **ii** 0.47 N **iii** 0.25
  iv 14° **v** 45°

SKILLS CHECK 6A (page 58)

1 **a** 8 N s **b** 1.8 N s **c** 10 N s
2 **a** 96 N s **b** 128 N s
3 4.67 m s^{-1}
4 0.75 m s^{-1}
5 26 m s^{-1}
6 3.69 m s^{-1}, 1850 N s
7 300 m s^{-1}, 45 000 N
8 6 m s^{-1}

Exam Practice 6 (page 58)

1 0.7 m s^{-1}
2 2000
3 **i** 3 m s^{-1} **ii** 0.4
4 **i** 1.8 m s^{-1} **ii** 0.7
5 **i** 1.5 m s^{-1} **ii** 0.21 m s^{-1}
6 **i** 6.3 m s^{-1} **ii** 0.025 **iii** 2.24 N s
7 **ii** 3.9 s **iii** 1.5 m s^{-1} **iv** 8.4 m
8 0.1, u m s^{-1}
9 **i** 0.15 **ii** $\frac{1}{15}, \frac{2}{15}, \frac{2}{5}$

Practice exam paper (page 61)

1 14 N s
2 105 m, 25 m s^{-1}
3 7.71
4 **a** 10 **b** 7 **c** 5.76
5 **a** 10 s
  b -6 m s^{-1}, midway between A and B, running towards A
  c 30 m
  d 1.2 m s^{-2}
  e Peaks and troughs closer to t-axis, t-intercepts further apart.
6 **b** 1.7 N, 0.218
  c P will slide back down the slope.
7 **a** 1.6 m s^{-1} away from C **b** 3.2 kg **d** 8.8 m
8 **a** 64 m s^{-1}, 40 s
  b -0.32 m s^{-2}, negative sign means aircraft is slowing down
  c 792 m

SINGLE USER LICENCE AGREEMENT FOR MECHANICS 1 FOR OCR CD-ROM

IMPORTANT: READ CAREFULLY

WARNING: BY OPENING THE PACKAGE YOU AGREE TO BE BOUND BY THE TERMS OF THE LICENCE AGREEMENT BELOW.

This is a legally binding agreement between You (the user or purchaser) and Pearson Education Limited. By retaining this licence, any software media or accompanying written materials or carrying out any of the permitted activities You agree to be bound by the terms of the licence agreement below.

If You do not agree to these terms then promptly return the entire publication (this licence and all software, written materials, packaging and any other components received with it) with Your sales receipt to Your supplier for a full refund.

YOU ARE PERMITTED TO:

- Use (load into temporary memory or permanent storage) a single copy of the software on only one computer at a time. If this computer is linked to a network then the software may only be used in a manner such that it is not accessible to other machines on the network.

- Transfer the software from one computer to another provided that you only use it on one computer at a time.

- Print a single copy of any PDF file from the CD-ROM for the sole use of the user.

YOU MAY NOT:

- Rent or lease the software or any part of the publication.

- Copy any part of the documentation, except where specifically indicated otherwise.

- Make copies of the software, other than for backup purposes.

- Reverse engineer, decompile or disassemble the software.

- Use the software on more than one computer at a time.

- Install the software on any networked computer in a way that could allow access to it from more than one machine on the network.

- Use the software in any way not specified above without the prior written consent of Pearson Education Limited.

- Print off multiple copies of any PDF file.

ONE COPY ONLY

This licence is for a single user copy of the software

PEARSON EDUCATION LIMITED RESERVES THE RIGHT TO TERMINATE THIS LICENCE BY WRITTEN NOTICE AND TO TAKE ACTION TO RECOVER ANY DAMAGES SUFFERED BY PEARSON EDUCATION LIMITED IF YOU BREACH ANY PROVISION OF THIS AGREEMENT.

Pearson Education Limited and/or its licensors own the software.
You only own the disk on which the software is supplied.

Pearson Education Limited warrants that the diskette or CD-ROM on which the software is supplied is free from defects in materials and workmanship under normal use for ninety (90) days from the date You receive it. This warranty is limited to You and is not transferable. Pearson Education Limited does not warrant that the functions of the software meet Your requirements or that the media is compatible with any computer system on which it is used or that the operation of the software will be unlimited or error free.

You assume responsibility for selecting the software to achieve Your intended results and for the installation of, the use of and the results obtained from the software. The entire liability of Pearson Education Limited and its suppliers and your only remedy shall be replacement free of charge of the components that do not meet this warranty.

This limited warranty is void if any damage has resulted from accident, abuse, misapplication, service or modification by someone other than Pearson Education Limited. In no event shall Pearson Education Limited or its suppliers be liable for any damages whatsoever arising out of installation of the software, even if advised of the possibility of such damages. Pearson Education Limited will not be liable for any loss or damage of any nature suffered by any party as a result of reliance upon or reproduction of or any errors in the content of the publication.

Pearson Education Limited does not limit its liability for death or personal injury caused by its negligence.

This licence agreement shall be governed by and interpreted and construed in accordance with English law.